Materials Forming, Machining and Tribology

Series Editor

J. Paulo Davim

For further volumes:
http://www.springer.com/series/11181

Bekir Sami Yilbas · Shahzada Zaman Shuja

Laser Surface Processing and Model Studies

Springer

Bekir Sami Yilbas
Shahzada Zaman Shuja
Mechanical Engineering Department
King Fahd University of Petroleum and Minerals
Dhahran
Saudi Arabia

ISSN 2195-0911 ISSN 2195-092X (electronic)
ISBN 978-3-642-36628-4 ISBN 978-3-642-36629-1 (eBook)
DOI 10.1007/978-3-642-36629-1
Springer Heidelberg New York Dordrecht London

Library of Congress Control Number: 2013932463

Printed on acid-free paper

Springer is part of Springer Science+Business Media (www.springer.com)

Acknowledgment

We would like to acknowledge the role of King Fahd University of Petroleum & Minerals in extending strong support from the beginning to the end facilitating every means during the preparation of this book. The authors wish to thank the colleagues who contributed to the work presented in the book through previous cooperation of the authors. In particular, thanks to Dr. Muammer Kalyon, Dr. Nasser Al-Aqeeli, Dr. Saad Bin Mansoor, and all my graduate students.

Contents

1 Introduction 1

2 Conduction Heating of Solid Surfaces 5
2.1 Introduction 5
2.2 Analytical Treatment of Laser Pulse Heating 6
2.2.1 Exponential Pulse Heating 6
2.2.2 Laser Repetitive Pulse Heating 11
2.3 Effect of Duty Cycle on Heating: Numerical Treatment 13
2.4 Discussions 15
2.4.1 Exponential Pulse Heating Case and Convection Condition Resembling Assisting Gas at the Surface 16
2.4.2 Repetitive Pulse Heating Case and Convection Condition Resembling Assisting Gas at the Surface 20
2.4.3 Effect of Duty Cycle on Heating and Convection Condition Resembling Assisting Gas at the Surface 24
References 27

3 Laser Melting of Solid Surfaces 29
3.1 Introduction 29
3.2 Analytical Treatment of Laser Melting Process 30
3.2.1 The Closed Form Solution 30
3.2.2 Influence of Assisting Gas on the Melt Layer Thickness 32
3.3 Numerical Treatment of Surface Melting 34
3.3.1 Stationary Heating Source 34
3.3.2 Influence of Pulse Profile on Temperature Field 39
3.3.3 Influence of Marangoni Flow on Temperature Field 41
3.3.4 Moving Heat Source Consideration 41

3.4 Discussions 43
3.4.1 Analytical Treatment of Laser Melting Process 43
3.4.2 Numerical Treatment of Laser Melting 44
3.4.3 Influence of Pulse Profile and Marangoni Flow on Temperature Field 48
3.4.4 Moving Heat Source Consideration 52
References 57

4 Laser Melting of Two Layer Materials 59
4.1 Introduction 59
4.2 Numerical Treatment of Melting Process 60
4.3 Discussions 62
4.3.1 Laser Melting of Two-Layer Structure Influence of Laser Pulse Repetition on Temperature and Flow Field in the Melt Pool 63
4.3.2 Influence of Coating Material on Temperature and Flow Field in the Melt Pool 69
4.3.3 Influence of Coating Material Thickness on Temperature and Flow Field in the Melt Pool 75
References 80

5 Laser Induced Evaporation at the Surface 81
5.1 Introduction 81
5.2 Analytical Treatment of Evaporation 82
5.3 Numerical Treatment of Evaporation 91
5.4 Numerical Solution of Governing Equations 99
5.4.1 Phase Change Process 100
5.4.2 Transiently Developing Vapor Jet 100
5.5 Discussions 101
5.5.1 Solution of Analytical Treatment of Evaporation 101
5.5.2 Predictions from Numerical Treatment of Evaporation 103
References 110

6 Practical Applications of Laser Surface Treatment 111
6.1 Introduction 111
6.2 Laser Shock Processing 112
6.2.1 Heating, Recoil Pressure, and Wave Analysis 113
6.2.2 Method of Solutions 116
6.2.3 Experimental 117
6.3 Laser Gas Assisted Nitriding 118
6.3.1 Thermal and Residual Stress Analysis 119
6.3.2 Experimental and Measurement of Young Modulus and Fracture Toughness 121

6.4 Laser Surface Treatment of Pre-prepared Alloy 123
6.5 Discussion 124
6.5.1 Laser Shock Processing of Steel Surface 124
6.5.2 Laser Gas Assisted Nitriding of Steel Surfaces 127
6.5.3 Laser Surface Treatment of Pre-prepared Ti Alloy 131
References 137

7 Concluding Remarks 139
7.1 Analytical Treatment for Melting and Evaporation Processes . . . 139
7.2 Numerical Treatment for Melting and Evaporation Processes . . . 140
7.3 Practical Applications 143

Index 145

Chapter 1
Introduction

Abstract Lasers are widely used in surface engineering because of their advantages over the conventional surface treatment methods. Some of these advantages include precision of operation, fast processing, and localized treatment. Laser surface treatment involves with solid heating and phase change at the irradiated surface. In this chapter, the importance of laser applications in surface engineering is introduced and some aspects of laser surface treatment are presented.

Lasers are widely used in surface engineering because of their advantages over the conventional surface treatment methods. Some of these advantages include precision of operation, fast processing, and localized treatment. Laser surface treatment involves with solid heating and phase change at the irradiated surface. The solid heating at the surface of the substrate material can be described through introducing laser conduction limited heating situation; in which case, the absorption of irradiated laser energy results in heat conduction only and it does not cause the phase change at the irradiated surface. When the laser beam interacts with a solid surface, electrons in the irradiated region absorb the incident energy and increase their excess energy in this region. This, in turn, results in thermal separation of the electron sub-system from the lattice sub-system. Electrons and lattice phonons thermally communicate to each other and electrons undergo several collisions with lattice phonons through which some of electron excess energy transfers to lattice phonons. Since the number of collisions and electron excess energy in the electron sub-system define the rate of energy transfer from the electron sub-system to the lattice sub-system, the duration of electron excess energy transfer becomes important for the type of energy transport in the solid substrate, such as equilibrium or non-equilibrium transport. Thermal equilibrium between electron and lattice sub-systems can be achieved when the interaction time becomes on the order of the thermalization time of the substrate material; therefore, for short interaction durations non-equilibrium energy transport governs the heating process. The equilibrium based Fourier heating law for heat conduction is not applicable to describe the non-equilibrium energy transfer in a solid substrate. In addition, the absorption depth of the incident radiation is small for

B. S. Yilbas and S. Z. Shuja, *Laser Surface Processing and Model Studies*,
Materials Forming, Machining and Tribology, DOI: 10.1007/978-3-642-36629-1_1,

metallic substrates and heat wave propagation at a finite speed takes place in the irradiated region. The Fourier heating law fails to predict the correct temperature rise in this region due to the consideration of infinite heat wave speed in the irradiated solid. The phase lagging in energy transfer takes place in the solid because of the short duration of heating and the small size of the absorption depth. In such heating situations, model studies associated with thermal wave propagation, such as telegraph equation, or two-temperature model, or electron kinetic theory approaches should be incorporated to describe the temperature field during the thermal loading. In general, most of the laser treatment processes associated with the surface engineering application are involved with the durations longer than the thermalization time of the substrate material and the use of the Fourier diffusion law becomes appropriate to describe the heating process.

Laser surface treatment of alloys provides dense layer with fine grains while improving the hardness of the surface through the phase change and rapid solidification in the surface region. Depending on the settings of laser output parameters, such as the duty cycle, the power intensity, the laser scanning speed, and the material properties, the depth of melt pool can be controlled. This is necessary for the practical laser surface treatment applications. Controlling the heating parameters becomes crucial to achieve the desired melt zone during the treatment process. The excessive laser irradiated power or low laser scanning speeds can cause evaporation at the surface, which increases the roughness of the treated surface while limiting the practical applications of the treated surface. Therefore, laser heating parameters should be selected properly in order to avoid poor quality of the treated surface and the excessive operational costs. In order to assess and select the proper levels of the laser surface treatment parameters, model studies of laser heating process including phase change and molten flow in the melt pool become necessary. In addition, model studies give physical insight into the process and provide useful information on the influence of melting parameters. The numerical and analytical predictions reduce the experimental cost and minimize the experimental time. In the case of irradiation of high intensity laser beams, the phase change occurs in the irradiated region and the energy transfer in the irradiated region can be formulated by introducing a non-conduction limited heating process. In this case, substrate material under goes solid heating, melting, and evaporation during the heating process. The duration to reach melting and evaporation is extremely short for the high power pulse heating operation. The rapid rise of temperature at the surface causes difficulties in experimenting accurately the laser-workpiece interactions. Consequently, model studies are easy to handle and give insight into the laser-workpiece interaction mechanism particularly for high intensity laser beams with short-pulse durations. During the irradiation pulse, the laser irradiated power is absorbed in the skin of the substrate surface resulting in volumetric heat source at the irradiated region. Although the depth of absorption is considerably shallow, it has a significant effect on the heat transport and the internal energy gain of the substrate material in this region. Therefore, when modeling the heating process in metallic substrates, the volumetric heat source due to the absorption of the incident radiation needs to be accounted in the analysis. In the case of the non-conduction limited heating situation, laser beam energy is absorbed in the surface region of the substrate material in accordance with the Fresnel law as well as by the surface vapor. Moreover, once the cavity is formed, the Fresnel absorption due to multiple reflections of the beam inside the cavity is enhanced. This process becomes visible at high power intensities of $\geq 10^{12}$ W/m^2, which is the laser power intensity at the workpiece surface. On the other hand, at machining intensities ($10^{10}-10^{11}$ W/m^2), the vapor front generated is small in size and laser beam absorption by the vapor front is insignificant. Consequently, the processing parameters, which control the energy input to the workpiece, influence the extent of the melted region and this should be determined with accuracy if the simulation is to give quantitative results.

During the melting process, a liquid phase is formed at the irradiated surface and a mush zone is developed in the neighborhood of the liquid zone. The mush zone consists of the liquid and the solid phases and the size of the mush zone depends on the rate of heat transfer from the melt pool and the thermal properties of the substrate material. In addition, the thickness of the melt layer is important for sound laser processing of surfaces, since the tribological properties of the surface can be improved significantly when thick melt layers are formed. In the melt layer, a Marangoni flow is developed due to the surface tension effect. The convection current developed in the melt pool influences the rate of heat transfer while modifying the melt pool size and the mushy zone thickness. This, in turn, influences the laser melting process and the resulting surface quality. In the model studies in relation to laser melting of the surfaces, the mushy zone consideration and the Marangoni flow should be incorporated. In addition, laser sintering of the surfaces is involved with laser controlled melting of two materials with different properties. Therefore, model studies should be extended to include proper boundary conditions with proper properties of the substrate materials for different laser treatment processes, such as sintering and quenching.

Chapter 2
Conduction Heating of Solid Surfaces

Abstract Laser conduction heating involves with a solid phase heating for a stationary or moving sources. In order to accommodate the absorption of irradiated laser energy, a volumetric source consideration should be incorporated. Since the analytical solution for such heating situation is possible, a closed form solution for the temperature field is provided firstly. In order to assess the influence of duty cycle on the temperature field, a numerical model is introduced. To study the thermal effects, two-dimensional axisymmetric solid is considered for a stationary source and three-dimensional heating situation is incorporated for a moving source in the model studies. The laser heating is involved with the assisting gas processing; therefore, the convection effect of the assisting gas is incorporated in the analysis. Since the heating duration is longer than the thermalization time of the substrate material for most of the surface treatment processes, the Fourier heating law is incorporated in the analysis. In this chapter, Laser conduction heating of solid surfaces is introduced and analytical approaches for temperature field in the irradiated region is presented for the appropriate boundary and heating conditions.

2.1 Introduction

Laser conduction heating involves with a solid phase heating for a stationary or moving sources. In order to accommodate the absorption of irradiated laser energy, a volumetric source consideration should be incorporated. Since the analytical solution for such heating situation is possible, a closed form solution for the temperature field is provided firstly. In order to assess the influence of duty cycle on the temperature field, a numerical model is introduced. To study the thermal effects, two-dimensional axisymmetric solid is considered for a stationary source and three-dimensional heating situation is incorporated for a moving source in the model studies. The laser heating is involved with the assisting gas processing; therefore, the convection effect of the assisting gas is incorporated in the analysis.

B. S. Yilbas and S. Z. Shuja, *Laser Surface Processing and Model Studies*,
Materials Forming, Machining and Tribology, DOI: 10.1007/978-3-642-36629-1_2,

Since the heating duration is longer than the thermalization time of the substrate material for most of the surface treatment processes, the Fourier heating law is incorporated in the analysis.

2.2 Analytical Treatment of Laser Pulse Heating

Closed form solution of the pulse laser heating of the solid surfaces enables to identify the affecting parameters through analytical expressions developed between the temperature field and the parameters. The closed form solution for laser repetitive pulse heating, therefore, becomes fruitful when examining the possibility of the steady heating at the surface by the laser repetitive pulses. Laser pulse heating of the solid substrates requires formulation of laser pulses due to the fact that laser intensity varies with time. This results in different heating situations due to temporal laser pulse behavior. In this case, two heating situations can be considered to account for the temporal behavior of the laser pulse intensity. The first case is involved with a single exponential pulse heating and the other case is the heating due to the multi-pulses (repetitive pulses). The closed form solutions for each cases is presented according to the below sub-headings.

2.2.1 Exponential Pulse Heating

The analytical formulation of laser pulse heating of solid substrate with a convective boundary condition at the surface is considered in the light of the previous study [1]. The time exponentially varying pulse intensity is incorporated in the analysis provided that two different pulse types are taken into account. In the first pulse type (half pulse), the intensity decays exponentially with time ($I_1 \exp(-\beta' \tau)$) while the intensity distribution resembling a typical actual laser pulse $I_1[\exp(-\beta' \tau) - \exp(-\gamma' \tau)]$ is considered in the second pulse type (full pulse). The closed-form solution obtained from the present study is compared with the previous formulations for the appropriate pulse and boundary conditions. The heating analysis is carried out for nanosecond laser pulses, which is longer than the thermalization time of the metal substrates; therefore, the Fourier theory of heating is used when modeling the heating process [2]. The heat transfer equation for a laser heating pulse can be written as:

$$\frac{\partial^2 T}{\partial x^2} + \frac{I_1 \delta}{k}\left(e^{-\beta t} - e^{-\gamma t}\right)e^{\delta x} = \frac{1}{\alpha}\frac{\partial T}{\partial t} \tag{2.1}$$

where I_1 is the power intensity, δ is the absorption depth, β and γ are the pulse parameters, and α is the thermal diffusivity.

The output pulse from a laser can be formulated through subtracting two exponential functions. Hence, the power intensity distribution of time exponentially varying pulse can be written as:

$$I = I_1 \left(e^{-\beta t} - e^{-\gamma t} \right) \tag{2.2}$$

where $I_1 = (1 - r_f)I_0$ and r_f is the reflection coefficient and I_o is the peak power intensity, and parameters β and γ can be chosen to give the appropriate rise time for the pulse. Since the governing equation of heat transfer is linear (Eq. 2.1) the closed form solution can be obtained for a half pulse first, then, the complete solution can be achieved by subtraction of the solutions for the individual parts of the time exponential pulse (half pulse). It should be noted that for the solution of full pulse, the ambient temperature is considered as zero $(T_0 = 0)$. This is necessary since the full pulse solution satisfies the convective boundary condition when $T_0 = 0$. The heat transfer equation for the half pulse becomes:

$$\frac{\partial^2 T}{\partial x^2} + \frac{I_1 \delta}{k} e^{-(\beta t + \delta x)} = \frac{1}{\alpha} \frac{\partial T}{\partial t} \tag{2.3}$$

The absorption depth of the substrate material is considerable smaller than the thickness of the substrate material. This allows one to consider the semi-infinite solid body, initially at uniform temperature, with convective boundary condition at the surface. It should be noted that convective boundary condition at the surface resembles the assisting gas jet effect at the surface of the solid substrate during the heating process. Therefore, the initial and boundary conditions are:

at time t $= 0$ $\quad T(x, 0) = 0$

at the surface $x = 0$ $\quad \left[\frac{\partial T}{\partial x}\right]_{x=0} = \frac{h}{k}(T(0,t) - T_o)$

and at $x = \infty$ $\quad T(t, \infty) = 0$

The solution of Eq. 2.3 can be obtained possibly through Laplace transformation method, i.e., with respect to t, the Laplace transformation of Eq. 2.3 yields:

$$\frac{\partial^2 \overline{T}}{\partial x^2} + \frac{I_1 \delta}{k} e^{-\delta x} \frac{1}{p + \beta} = \frac{1}{\alpha} \left[p\overline{T} - T(x, 0) \right] \tag{2.4}$$

where $\overline{T} = T(x, p)$ and $T(x, 0) = 0$ due to the initial condition. Using the initial condition, Eq. 2.4 yields:

$$\frac{\partial^2 \overline{T}}{\partial x^2} - \frac{p\overline{T}}{\alpha} = -\frac{I_1 \delta}{k} e^{-\delta x} \frac{1}{p + \beta} \tag{2.5}$$

Let consider $\lambda^2 = \frac{p}{\alpha}$ and $H_o = -\frac{I_1 \delta}{k} \frac{1}{(p+\beta)}$, then Eq. 2.5 becomes:

$$\frac{\partial^2 \overline{T}}{\partial x^2} - \lambda^2 \overline{T} = H_o e^{-\delta x} \tag{2.6}$$

Equation 2.6 has homogeneous $(\overline{T}_h)$ and particular solutions $(\overline{T}_p)$, i.e.:

$$\overline{T} = \overline{T}_h + \overline{T}_p \tag{2.7}$$

The homogeneous solution is:

$$\overline{T}_h = C_1 e^{\lambda x} + C_2 e^{-\lambda x} \tag{2.8}$$

where C_1 and C_2 are the constants to be determined from the boundary conditions. Similarly the particular solution is:

$$\overline{T}_p = A_o e^{-\delta x} \tag{2.9}$$

where A_o is the constant.

Equation 2.7 yields the solution:

$$A_o \delta^2 e^{-\delta x} - \lambda^2 A_o e^{-\delta x} = H_o e^{-\delta x} \tag{2.10}$$

or

$$A_o = \frac{H_o}{(\delta^2 - \lambda^2)} \tag{2.11}$$

After the rearrangement, the particular solution $(\overline{T}_p)$ results:

$$\overline{T}_p = \frac{H_o}{(\delta^2 - \lambda^2)} e^{-\delta x} \tag{2.12}$$

Therefore, the solution of Eq. 2.4 in the Laplace domain becomes:

$$\overline{T} = \overline{T}(x,p) = C_1 e^{\lambda x} + C_2 e^{-\lambda x} + \frac{H_o}{(\delta^2 - \lambda^2)} e^{-\delta x} \tag{2.13}$$

The coefficients in Eq. 2.13 can be obtained from the boundary conditions. Consider $\lambda = \sqrt{\frac{p}{\alpha}} > 0$ and from the boundary condition $T(\infty, t) = 0$ result in $C_1 = 0$. Moreover, let $\overline{H}_0 = \frac{H_0}{(\delta^2 - \lambda^2)}$, Eq. 2.13 becomes:

$$\overline{T} = C_2 e^{-\lambda x} + \overline{H}_o e^{-\delta x} \tag{2.14}$$

In order to determine C_2, the boundary condition at the surface can be incorporated, i.e.:

$$\left[\frac{\partial \overline{T}}{\partial x}\right]_{x=0} = \frac{h}{k}\left[\overline{T}(0,p) - \frac{T_o}{p}\right] \tag{2.15}$$

where T_0, which is an ambient temperature (same as the initial temperature), can be specified. Introducing Eq. 2.14 into Eq. 2.15 and knowing that $\overline{T}(0,p) = C_2 + \overline{H}_o$, it yields:

$$-\lambda C_2 - \delta \overline{H}_o = \frac{h}{k}\left[C_2 + \overline{H}_o - \frac{T_o}{p}\right] \tag{2.16}$$

Hence, C_2 becomes:

$$C_2 = -\frac{\overline{H}_o(h + k\delta)}{(h + k\lambda)} + \frac{T_o h}{p(h + k\lambda)} \tag{2.17}$$

Substituting C_2 and the values of H_o, $\overline{H}_o$ and λ into Eq. 2.14, it becomes:

$$\overline{T}(x,p) = \frac{I_1\delta(h + k\delta)e^{-\sqrt{\frac{p}{\alpha}}x}}{k(p + \beta)\left(\delta^2 - \frac{p}{\alpha}\right)\left(h + \frac{k\sqrt{p}}{\sqrt{\alpha}}\right)} + \frac{T_o h e^{-\sqrt{\frac{p}{\alpha}}x}}{p\left(h + \frac{k\sqrt{p}}{\sqrt{\alpha}}\right)} - \frac{I_1\delta}{k(p + \beta)\left(\delta^2 - \frac{p}{\alpha}\right)}e^{-\delta x} \tag{2.18}$$

The mathematical arrangements of inversion of Eq. 2.18 is given in the previous study [1]. Therefore, the Laplace inversion of Eq. 2.18 becomes:

$$\begin{aligned} T(x,\, t) = a_{10}&\left\{\begin{array}{c} \sqrt{\beta}ie^{-\beta t}\left[-l_1 e^{\sqrt{\frac{\beta}{\alpha}}ix}\mathrm{erf\,c}\left(\sqrt{\beta t}i + \frac{x}{2\sqrt{\alpha t}}\right) + l_2 e^{-\sqrt{\frac{\beta}{\alpha}}ix}\mathrm{erf\,c}\left(-\sqrt{\beta t}i + \frac{x}{2\sqrt{\alpha t}}\right)\right] \\ +\delta\sqrt{\alpha}e^{\alpha\delta^2 t}\left[l_3 e^{-\delta x}\mathrm{erf\,c}\left(-\delta\sqrt{\alpha t} + \frac{x}{2\sqrt{\alpha t}}\right) - l_4 e^{\delta x}\mathrm{erf\,c}\left(\delta\sqrt{\alpha t} + \frac{x}{2\sqrt{\alpha t}}\right)\right] \\ -w_1 l_5 e^{w_1\frac{x}{\sqrt{\alpha}}}e^{w_1^2 t}\mathrm{erf\,c}\left(w_1\sqrt{t} + \frac{x}{2\sqrt{\alpha t}}\right) \end{array}\right\} \\ &+ \frac{a_{20}}{(\beta + \alpha\delta^2)}\left(e^{\alpha\delta^2 t} - e^{-\beta t}\right)e^{-\delta x} \\ &+ \frac{a_{30}}{w_1}\left[-e^{w_1\frac{x}{\sqrt{\alpha}}}e^{w_1^2 t}\mathrm{erf\,c}\left(w_1\sqrt{t} + \frac{x}{2\sqrt{\alpha t}}\right) + \mathrm{erf\,c}\left(\frac{x}{2\sqrt{\alpha t}}\right)\right] \end{aligned} \tag{2.19}$$

where

$$\begin{aligned} l_1 &= \frac{1}{2\sqrt{\beta}i(\beta + \alpha\delta^2)(-\sqrt{\beta}i + w_1)} : l_2 = -\frac{1}{2\sqrt{\beta}i(\beta + \alpha\delta^2)(\sqrt{\beta}i + w_1)} \\ l_3 &= \frac{1}{2\sqrt{\alpha}\delta(\beta + \alpha\delta^2)(\delta\sqrt{\alpha} + w_1)} : l_4 = -\frac{1}{2\sqrt{\alpha}\delta(\beta + \alpha\delta^2)(-\delta\sqrt{\alpha} + w_1)} \\ l_5 &= \frac{1}{(w_1^2 + \beta)(w_1^2 - \alpha\delta^2)} \end{aligned} \tag{2.20}$$

and

$$a_{10} = -\frac{I_1\alpha\sqrt{\alpha}\delta(h + k\delta)}{k^2} : a_{20} = \frac{I_1\alpha\delta}{k} : a_{30} = \frac{h\sqrt{\alpha}}{k}T_o : w_1 = \frac{h\sqrt{\alpha}}{k} \tag{2.21}$$

After inserting l_1, l_2, l_3, l_4, l_5, a_{10}, a_{20}, a_{30}, and w_1, Eq. 2.19 becomes:

$$T(x,t) = \frac{I_1 \alpha \sqrt{\alpha}\delta(h+k\delta)}{k^2} \left\{ \begin{array}{l} -\frac{e^{-\delta x} e^{\alpha\delta^2 t} \operatorname{erf} c\left(\frac{x}{2\sqrt{\alpha t}} - \delta\sqrt{\alpha t}\right)}{2(\beta+\alpha\delta^2)\left(\frac{h\sqrt{\alpha}}{k} + \delta\sqrt{\alpha}\right)} - \frac{e^{\delta x} e^{\alpha\delta^2 t} \operatorname{erf} c\left(\frac{x}{2\sqrt{\alpha t}} + \delta\sqrt{\alpha t}\right)}{2(\beta+\alpha\delta^2)\left(\frac{h\sqrt{\alpha}}{k} - \delta\sqrt{\alpha}\right)} \\ + \frac{e^{-\beta t} e^{\sqrt{\frac{\beta}{\alpha}} x i} \operatorname{erf} c\left(\frac{x}{2\sqrt{\alpha t}} + \sqrt{\beta t i}\right)}{2(\beta+\alpha\delta^2)\left(\frac{h\sqrt{\alpha}}{k} - \sqrt{\beta i}\right)} + \frac{e^{-\beta t} e^{-\sqrt{\frac{\beta}{\alpha}} x i} \operatorname{erf} c\left(\frac{x}{2\sqrt{\alpha t}} - \sqrt{\beta t i}\right)}{2(\beta+\alpha\delta^2)\left(\frac{h\sqrt{\alpha}}{k} + \sqrt{\beta i}\right)} \\ + \frac{h\sqrt{\alpha}}{k} \frac{e^{\frac{hx}{k}} e^{\frac{h^2}{k^2}\alpha t} \operatorname{erf} c\left(\frac{x}{2\sqrt{\alpha t}} + \frac{h}{k}\sqrt{\alpha t}\right)}{k\left(\frac{h^2\alpha}{k^2} + \beta\right)\left(\frac{h^2\alpha}{k^2} - \alpha\delta^2\right)} - \frac{k\sqrt{\alpha} e^{-\delta x}\left(e^{-\alpha\delta^2 t} - e^{-\beta x}\right)}{(\beta+\alpha\delta^2)(h+k\delta)} \end{array} \right\}$$

$$+ T_o\left[\operatorname{erf} c\left(\frac{x}{2\sqrt{\alpha t}}\right) - e^{\frac{hx}{k}} e^{\frac{h^2}{k^2}\alpha t} \operatorname{erf} c\left(\frac{x}{2\sqrt{\alpha t}} + \frac{h}{k}\sqrt{\alpha t}\right)\right] \quad (2.22)$$

The surface temperature can be obtained when setting x equals to zero ($x = 0$) in Eq. 2.22. The mathematical arrangements are given in the previous study [1]. Therefore, the resulting surface temperature becomes:

$$T(0.t) = \frac{I_1 \alpha\delta}{(\beta+\alpha\delta^2)(\beta k^2 + \alpha h^2)}\left[(\alpha\delta h - \beta k)e^{-\beta t} + 2\sqrt{\frac{\alpha\beta}{\pi}}(h+\delta k)F(\sqrt{\beta t})\right]$$
$$+ \frac{I_1 \alpha\delta^2}{(\delta k - h)(\beta+\alpha\delta^2)} e^{\alpha\delta^2 t} \operatorname{erf} c(\delta\sqrt{\alpha t})$$
$$- \left(T_o - \frac{I_1 \alpha\delta h k}{(\beta k^2 + \alpha h^2)(h - \delta k)}\right) e^{\frac{h^2 \alpha t}{k^2}} \operatorname{erf} c\left(\frac{h}{k}\sqrt{\alpha t}\right) + T_o \quad (2.23)$$

In order to reduce Eq. 2.23 in dimensionless form, the following non-dimensional parameters are introduced:

$$\beta' = \frac{\beta}{\alpha\delta^2} : h' = \frac{h}{\delta k} : \tau = \alpha\delta^2 t \quad (2.24)$$

After mathematical arrangement and incorporating Eqs. 2.24, 2.23 yields:

$$T(0,\tau) = \frac{I_1}{\delta k}\frac{1}{(1+\beta')(\beta' + h'^2)}\left[(h' - \beta')e^{-\beta'\tau} + \frac{2}{\sqrt{\pi}}\sqrt{\beta'}(1+h')F\left(\sqrt{\beta'\tau}\right)\right]$$
$$+ \frac{I_1}{\delta k}\frac{1}{(1-h')(1+\beta')} e^{\tau} \operatorname{erf} \mathrm{c}(\sqrt{\tau})$$
$$- \left[T_o - \frac{I_1}{\delta k}\left(\frac{h'}{(\beta' + h'^2)(h' - 1)}\right)\right] e^{h'^2 \tau} \operatorname{erf} \mathrm{c}(h'\sqrt{\tau}) + T_o \quad (2.25)$$

where $F(\chi)$ is Dawson's integral, which is:

$$F(\chi) = e^{-\chi^2} \int_0^{\chi} e^{\xi^2} d\xi \tag{2.26}$$

The complete solution for the heating pulse (full pulse) can be obtained through incorporating the principle of superposition; in which case, after subtracting the solutions of the temperature profile for two exponential terms $\left(e^{-\beta t} \text{ and } e^{-\gamma t}\right)$ as used in the full pulse profile, the complete solution for the full pulse becomes:

$$\begin{aligned}
T(0,\tau) = & \frac{I_1}{\delta k} \frac{1}{(1+\beta')(\beta'+h'^2)} \left[(h'-\beta')e^{-\beta'\tau} + \frac{2}{\sqrt{\pi}} \sqrt{\beta'}(1+h')F\left(\sqrt{\beta'\tau}\right) \right] \\
& + \frac{I_1}{\delta k} \frac{1}{(1-h')(1+\beta')} e^{\tau} \operatorname{erf} c(\sqrt{\tau}) \\
& + \frac{I_1}{\delta k} \frac{h'}{(\beta'+h'^2)(1-h')} e^{h'^2\tau} \operatorname{erf} c(h'\sqrt{\tau}) \\
& \frac{I_1}{\delta k} \frac{1}{(1+\gamma')(\gamma'+h'^2)} \left[(h'-\gamma')e^{-\gamma'\tau} + \frac{2}{\sqrt{\pi}} \sqrt{\gamma'}(1+h')F(\gamma'\tau) \right] \\
& + \frac{I_1}{\delta k} \frac{1}{(1-h')(1+\gamma')} e^{\tau} \operatorname{erf} c(\sqrt{\tau}) \\
& + \frac{I_1}{\delta k} \frac{h'}{(\gamma'+h'^2)(1-h')} e^{h'^2\tau} \operatorname{erf} c(h'\sqrt{\tau})
\end{aligned} \tag{2.27}$$

where $\gamma' = \frac{\gamma}{\alpha\delta^2}$. Equation 2.27 satisfies the convection boundary condition for zero ambient temperature ($T_o = 0$). Equations 2.25 and 2.27 can be used to compute the dimensionless surface temperature profiles at the surface for half and full laser heating pulses.

2.2.2 *Laser Repetitive Pulse Heating*

Laser repetitive pulse heating of the solid surface with convective boundary condition at the surface is considered to account for the assisting gas jet effect. A closed form solution for laser heating process is obtained using a Laplace transformation method in line with the previous study [1]. The conditions for constant temperature heating at the surface are investigated and the pulse parameter $\left(\frac{\beta_0}{\gamma_0}\right)$ resulting in possible steady temperature attainment at the surface is discussed.

In order to account for the repetitive pulse heating, the intensity profile resembling the consecutive pulses should be incorporated [3]. This can be achieved introducing a time shift in the intensity function, which is given in

Eq. 2.2. Therefore, the heat transfer equation employing the consecutive pulses can be written as [3]:

$$\frac{\partial^2 T}{\partial x^2} + \frac{I_1 \delta}{k}\left(e^{-\beta t} - e^{-\gamma t}\right) e^{\delta x} F(t - t_o) = \frac{1}{\alpha}\frac{\partial T}{\partial t} \tag{2.28}$$

where $F(t - t_o)$ is a step function, which aids to resemble the consecutive pulses, i.e.:

$$F(t - t_o) = \left\{ \begin{matrix} 0, \; t < t_o \\ 1, \; t \geq t_o \end{matrix} \right\} \tag{2.29}$$

The solution of Eq. 2.28 is identical to Eq. 2.22 provided that the time term is replaced by $(\tau - \tau_o)$ and the solution is multiplied by a non-dimensional step function. The solution of Eq. 2.28 for the surface temperature yields:

$$T(0,\tau) = F(\tau - \tau_o)\left\{ \begin{aligned} & \frac{I_1}{\delta k}\frac{1}{(1+\beta_o)(\beta_o+h_o^2)}\left[\begin{matrix} (h_o - \beta_o)e^{-\beta_o(\tau-\tau_o)} \\ +\frac{2}{\sqrt{\pi}}\sqrt{\beta_o}(1+h_o)F\left(\sqrt{\beta_o(\tau-\tau_o)}\right) \end{matrix} \right] \\ & + \frac{I_1}{\delta k}\frac{1}{(1-h_o)(1+\beta_o)}e^{\tau-\tau_o}\operatorname{erf}c\left(\sqrt{\tau-\tau_o}\right) \\ & + \frac{I_1}{\delta k}\frac{h_o}{(\beta_o+h_o^2)(1-h_o)}e^{h_o^2(\tau-\tau_o)}\operatorname{erf}c\left(h_o\sqrt{\tau-\tau_o}\right) \\ & + \frac{I_1}{\delta k}\frac{1}{(1+\gamma_o)(\gamma_o+h_o^2)}\left[\begin{matrix} (h_o - \gamma_o)e^{-\gamma_o(\tau-\tau_o)} \\ +\frac{2}{\sqrt{\pi}}\sqrt{\gamma_o}(1+h_o)F\left(\gamma_o(\tau-\tau_o)\right) \end{matrix} \right] \\ & + \frac{I_1}{\delta k}\frac{1}{(1-h_o)(1+\gamma_o)}e^{\tau-\tau_o}\operatorname{erf}c\left(\sqrt{\tau-\tau_o}\right) \\ & + \frac{I_1}{\delta k}\frac{h_o}{(\gamma_o+h_o^2)(1-h_o)}e^{h_o^2(\tau-\tau_o)}\operatorname{erf}c\left(h_o\sqrt{\tau-\tau_o}\right) \end{aligned} \right\} \tag{2.30}$$

where $F(\tau - \tau_o) = \left\{ \begin{matrix} 0, \; \tau < \tau_o \\ 1, \; \tau \geq \tau_o \end{matrix} \right\}$ and $\tau_o = \alpha\delta^2 t_o$.

Equation 2.30 is used to compute the dimensionless surface temperature profiles at the surface for a complete laser heating pulse.

In order to determine the peak temperature differences and the maximum temperature difference in the surface temperature profile due to repetitive pulses, the followings are introduced:

$$\Delta T_{p1} = T_{p1} - T_{p2} \tag{2.31}$$

where ΔT_{p1} is the first peak temperature difference, T_{p1} and T_{p2} are the peak surface temperatures corresponding to the first and second consecutive pulses, respectively. The second peak temperature difference can be written as:

$$\Delta T_{p2} = T_{p2} - T_{p3} \tag{2.32}$$

where T_{p3} is the peak surface temperatures corresponding to the third consecutive pulse. The maximum temperature difference in between the first and second peaks of the temperature profile is:

$$\Delta T_{\max 1} = T_{p1} - T_{\min 1} \tag{2.33}$$

where $T_{\min 1}$ is the minimum surface temperature in between the first and second consecutive pulses. The maximum temperature difference in between the second and third peaks of the temperature profile is:

$$\Delta T_{\max 2} = T_{p2} - T_{\min 2} \tag{2.34}$$

where $T_{\min 2}$ is the minimum surface temperature in between the second and third consecutive pulses.

2.3 Effect of Duty Cycle on Heating: Numerical Treatment

In practical laser heating situations, laser pulses employed have single duty cycle (energy in per pulse or energy content of a single pulse) and cycle frequency. It has been demonstrated that the repetition rate of laser pulses improves the laser processing efficiency. However, the laser pulses, in general, have a rise and a fall times, which make the analytical solution difficult to achieve, particularly for two-dimensional axisymmetric heating situations. Therefore, numerical solution for such heating situation becomes unavoidable. Since the heating process is transient, initial and boundary conditions become important resembling the actual physical process.

Consider the laser repetitive pulse heating of solid surface, the governing equation of heat diffusion consistent with the conduction limited heating situation can be written for a two-dimensional semi-infinite solid. Laser beam power intensity distribution at the solid surface can be assumed to be Gaussian with its spot center at the center of the coordinate system. The transient heat conduction equation for a solid substrate irradiated by a laser beam with a Gaussian intensity profile at the surface can be written as:

$$\rho Cp \frac{\partial T}{\partial t} = k\left[\frac{1}{r}\frac{\partial}{\partial r}\left(r\frac{\partial T}{\partial r}\right) + \frac{\partial^2 T}{\partial x^2}\right] + S_0 \tag{2.35}$$

where x is the axial direction (along the laser beam axis), r is the radial direction (normal to the laser beam axis) ρ is the density, k is the thermal conductivity, Cp is the specific heat and S_0 is the volumetric source term. The volumetric heat source can be arranged to resemble the laser repetitive pulses; therefore,

$$S_0 = I_0 f(t)\delta(1 - r_f)\exp(-\delta x)\exp\left(-\frac{r^2}{a^2}\right) \tag{2.36}$$

where I_0, δ, r_f and a are the power intensity, absorption depth and the Gaussian parameter, respectively.

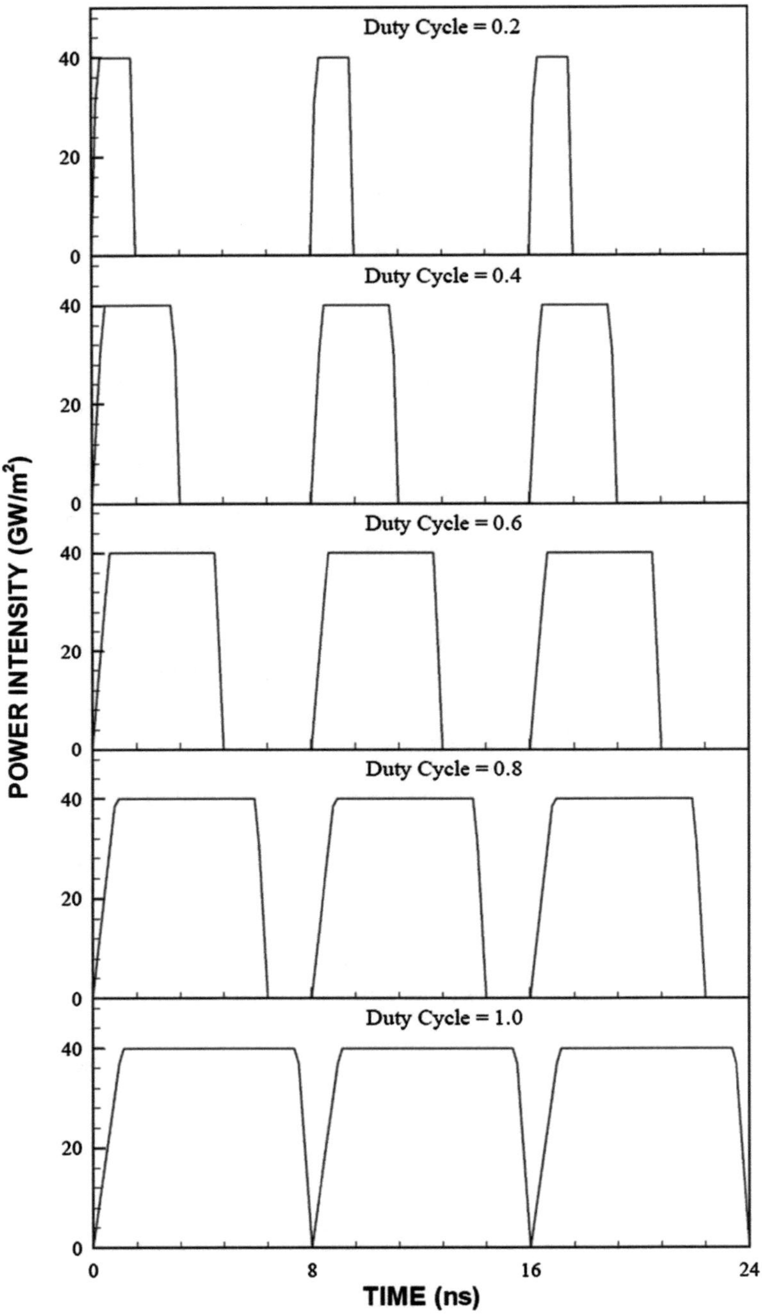

Fig. 2.1 Laser pulses with five duty cycles for the first three consecutive pulses

The temporal variation of pulse intensity resembling laser repetitive pulses can be defined by the function $f(t)$, which is:

$$f(t) = \begin{Bmatrix} 0 \le t \le t_r & at \\ -t_r < t \le t_p & 1 \\ t_p < t \le t_f & -bt \\ t_f < t < t_c & 0 \end{Bmatrix} \tag{2.37}$$

where t_r and t_f are the constants defining the rise and fall of the laser pulse, t_p is the pulse length, and a and b are the constants. The pulse shape used in the simulation is shown in Fig. 2.1.

The initial condition can be represented as the substrate material is assumed to be initially ($t = 0$) at a uniform temperature T_0 which can be specified.

In order to represent the assisting gas, a convective boundary condition is assumed at the surface ($x = 0$). The heat transfer coefficient used in the present simulations is $h = 10^4$ W/m^2K [1]. Therefore, at the surface, where $x = 0$, the boundary condition yields $\frac{\partial T}{\partial x} = h(T_s - T_0)$.

At the symmetry axis ($r = 0$) temperature is assumed to be maximum. Therefore, at the symmetry axis ($r = 0$), the boundary condition is $\frac{\partial T}{\partial r} = 0$. At a distance far away from the surface distance significantly larger than the absorption depth of the substrate material (δ), a room temperature is assumed, i.e. as $x \to \infty$ $T = T_0$.

Equation 2.35 is simulated numerically to obtain thermal response of the substrate material due to the repetitive heating pulse.

The numerical scheme employing an implicit formulation is used. To discretize the governing equations, a control volume approach is introduced. The calculation domain is divided into grids and grid independence test is being performed for different grid size and orientation. A non-uniform grid with 95 $\times$ 80 mesh points along x and r axes, respectively, is employed after securing the grid independence. The finer grids are located near the irradiated spot center in the vicinity of the surface and grids become courser as the distance increases towards the bulk of substrate material. A computer program based on implicit scheme is used to compute the temperature field.

2.4 Discussions

The comparison of the analytical solutions with the literature [4, 5] and temperature predictions from the analytical solutions are presented in line with the previous studies [1, 2, 6]. The findings in relation to analytical solutions and the numerical predictions are discussed according to the sub-headings given below.

2.4.1 *Exponential Pulse Heating Case and Convection Condition Resembling Assisting Gas at the Surface*

The closed-form solution for the temperature profiles due to time exponentially varying pulse is obtained. In order to account for the pulse resembling a typical actual pulse, two different time exponentially varying pulse intensities are considered. The heating involved with the time exponentially varying pulse, having a single time exponential term, is called the half pulse case while for the time exponentially varying pulse, having two time exponential terms, is called the full pulse case in the discussion section. The closed-form solution obtained is compared with the analytical solutions reported in the literature for different heating conditions [4, 6]. The data used in the solutions are given in Table 2.1.

To deduce the closed-form solution derived from the present analysis to the analytical solutions reported in the literature, the following heating conditions are considered: (1) step input intensity pulse with convective boundary condition at the surface, and (2) time exponentially varying pulse with non-convective boundary condition at the surface. The analytical solution for the step input intensity pulse with convective boundary condition was obtained previously by Blackwell [4]. In order to introduce the step input intensity pulse in the closed-form solution derived at present, the pulse parameters (β' and γ') in the source term of Eq. 2.2 or in the closed-form solution Eq. 2.25 should be set to zero. Moreover, the initial and ambient temperatures, and the symbol of absorption coefficient used in Blackwell's solution are different than those employed in the present analysis, i.e. the initial temperature is set to zero, the ambient temperature is denoted as T_0 and δ is used for the absorption coefficient in the present study. The resulting equation is:

$$
\begin{aligned}
T(x,t) = & T_o\left[\operatorname{erfc}\left(\frac{x}{2\sqrt{\alpha t}}\right) - e^{\left(\frac{h^2}{k^2}\alpha t+\frac{h}{k}x\right)}\operatorname{erfc}\left(\frac{x}{2\sqrt{\alpha t}}+\frac{h}{k}\sqrt{\alpha t}\right)\right] \\
& + \frac{I_1}{k\delta}\left\{\begin{array}{c}
\left(1+\frac{k\delta}{h}\right)\operatorname{erfc}\left(\frac{x}{2\sqrt{\alpha t}}\right) - \frac{1}{2}e^{(\alpha\delta^2 t-\delta x)}\operatorname{erfc}\left(-\delta\sqrt{\alpha t}+\frac{x}{2\sqrt{\alpha t}}\right) \\
-\frac{1}{2}\frac{\left(\frac{h}{k\delta}+1\right)}{\left(\frac{h}{k\delta}-1\right)}e^{(\alpha\delta^2 t+\delta x)}\operatorname{erfc}\left(\delta\sqrt{\alpha t}+\frac{x}{2\sqrt{\alpha t}}\right) \\
+\frac{1}{\frac{h}{k\delta}\left(\frac{h}{k\delta}-1\right)}e^{\left(\frac{h}{k}x+\frac{h^2}{k^2}\alpha t\right)}\operatorname{erfc}\left(\frac{h}{k}\sqrt{\alpha t}+\frac{x}{2\sqrt{\alpha t}}\right)+e^{-\delta x}\left(e^{\alpha\delta^2 t}-1\right)
\end{array}\right\}
\end{aligned}
\tag{2.38}
$$

Equation 2.38 is identical to the analytical solution obtained by Blackwell [4].

Table 2.1 Data used in the solutions for exponential pulse heating

Substrate	$\delta \times 10^7$ (1/m)	$\alpha \times 10^{-4}$ (m^2/s)	c_p (J/kgK)	ρ (kg/m^3)	k (1/s)	γ (1/s)	I_0 (W/m^2)	
Steel	6.16	0.227	460	7,880	80.3	0.1–0.3	0.3	10^{11}

The analytical solution for the non-convective boundary condition at the surface and time exponentially varying intensity pulse was obtained by Yilbas [6] previously. To introduce the non-convective boundary condition at the surface, the heat transfer coefficient (h) in Eq. 2.25 is set to zero. In this case, for a half pulse, Eq. 2.25 reduces to:

$$T(0,\tau) = \frac{I_1}{k\delta(1+\beta')}\left[e^{\tau}\mathrm{erfc}(\sqrt{\tau}) + \frac{2}{\sqrt{\pi}}\frac{1}{\sqrt{\beta'}}F\left(\sqrt{\beta'\tau}\right) - e^{-\beta'\tau}\right] \tag{2.39}$$

Equation 2.39 is identical to the analytical solution obtained previously by Yilbas [6].

Figure 2.2 shows the temporal variation of the surface temperature for the half pulse case and constant heat transfer coefficient $\left(h = 10^4\ \mathrm{W/m^2}K\right)$ at the surface while pulse parameter (β') is variable. In general, the rise of the temperature profiles is similar during the initial period of heating. As β' reduces, temperature reaches its peak value earlier. This occurs because of the time variation of the heating pulse as depicted in Fig. 2.3. In this case, as β' increases, the power intensity distribution leans towards the pulse beginning provided that the area under the pulse intensity curve becomes less. Consequently, the energy content of the heating pulse reduces as β' increases, which in turn results in low peak temperature at the surface of the substrate. The rate of surface temperature rise and decay varies as β' varies. This can also be seen from Fig. 2.4, in which $\frac{\partial T}{\partial \tau}$ with time (τ) is shown. $\frac{\partial T}{\partial \tau}$ decays rapidly as τ increases from the pulse beginning. The slope of $\frac{\partial T}{\partial \tau}$ curve reduces to zero at time corresponding to the peak temperature. The duration at which rapid decay of $\frac{\partial T}{\partial \tau}$ indicates the rapid increase of internal energy gain of the substrate, which dominates over the conduction and convection energy transport. At the time of minimum $\frac{\partial T}{\partial \tau}$, internal energy gain of the substrate becomes considerably small, since the pulse energy corresponding to this heating

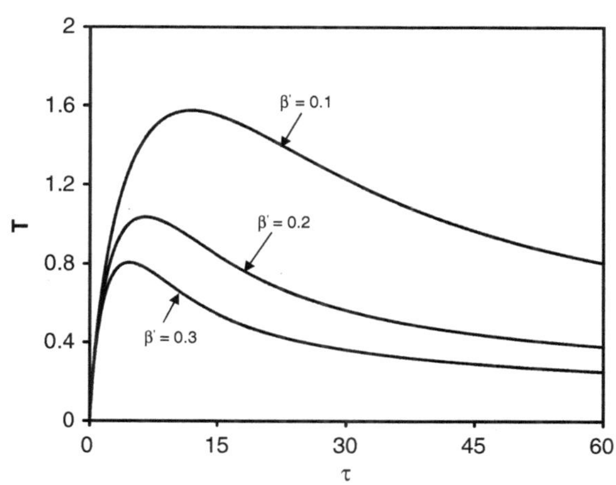

Fig. 2.2 Temporal variation of temperature for $\mathrm{Bi} = 2 \times 10^{-5}$ and β' is variable

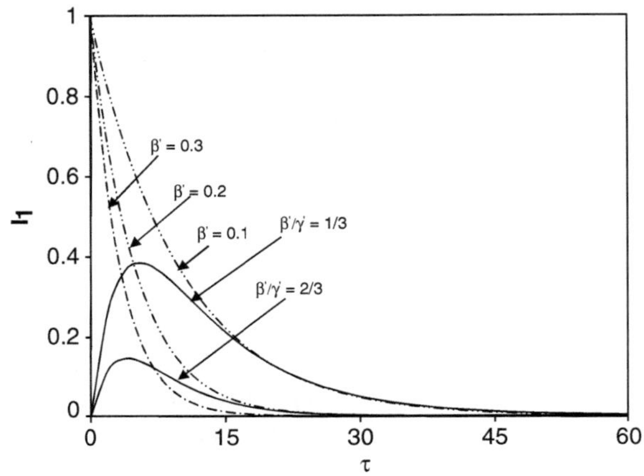

Fig. 2.3 Temporal variation of power intensity distribution as β', γ' and β'/γ' are variable

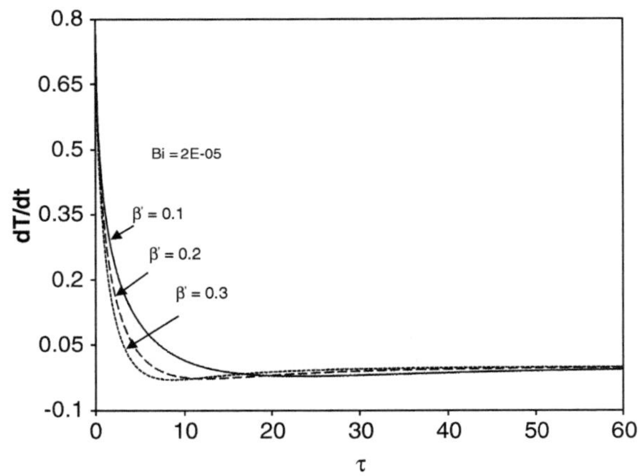

Fig. 2.4 Temporal variation of time gradient of temperature for $Bi = 2 \times 10^{-5}$ and different β' values

time is very small. As pulse intensity ceases, the conduction and convection become the sole mechanisms in energy transport process. In accordance with the previous study [7], the equilibrium time can be introduced at the point of minimum $\frac{\partial T}{\partial \tau}$. In this case, the energy balance attains among the internal energy gain, due to absorption of laser pulse, conduction, and convection contribution of the energy transport in the surface vicinity of the substrate. The equilibrium time shifts close to the pulse beginning as β' reduces. This is due to the temporal distribution of the laser pulse intensity as indicated earlier.

Figure 2.5 shows the temporal variation of surface temperature for different values of Biot number (B_i) and fixed β' value. The effect of Bi on the temperature profiles is not significant for $Bi \leq 0.02$. This because the energy absorbed by the substrate, which is considerably high as compared to the energy transported due to convection from the surface of the substrate. However, as Bi increases further, the

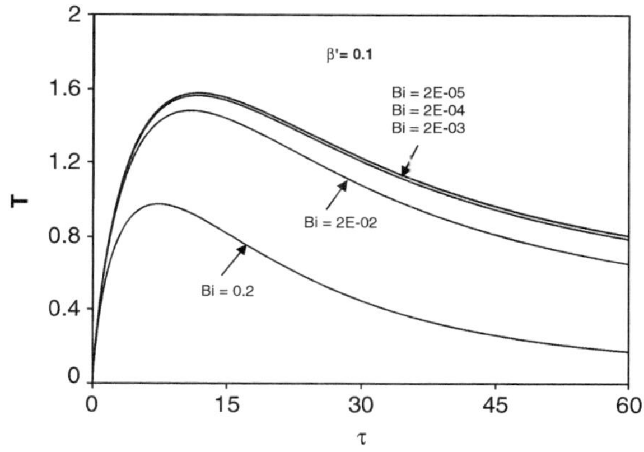

Fig. 2.5 Temporal variation of temperature for different $\beta' = 0.1$ and Biot number

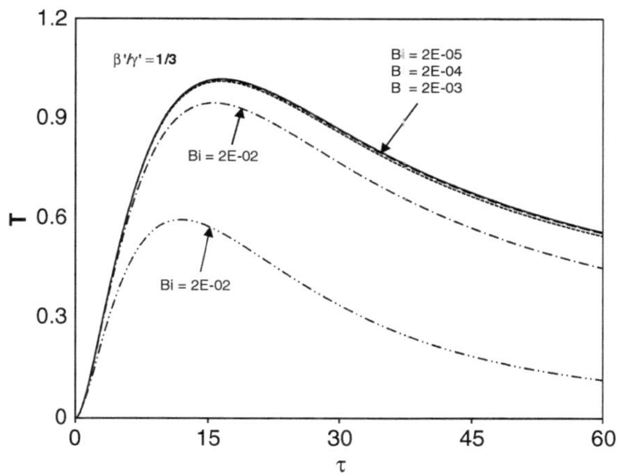

Fig. 2.6 Temporal variation of temperature for $\beta'/\gamma' = 1/3$ and Biot number is variable

rate of rise and the value of peak temperature at the surface reduce. In this case, energy transported from the surface because of convection becomes considerable. The time corresponding to peak temperature moves close to the pulse beginning for high *Bi*.

In order to resemble the actual laser pulse, a full pulse is considered; in which case, the pulse profile $I = I_1[\exp(-\beta'\tau) - \exp(-\gamma'\tau)]$ is employed in the source term of Eq. 2.1. Figure 2.6 shows the temporal variation of temperature profiles at the surface obtained for full pulse profile, in which $\beta' = 0.1$ and $\gamma' = 0.3$ are employed. The rate of change of temperature profile in the pulse beginning is lower than that corresponding to a half pulse for $\beta' = 0.1$. This is because the power intensity corresponding to a half pulse is higher than its counterpart corresponding to a full pulse Fig. 2.3. As similar to the behavior of *Bi* for the half pulse, the effect of *Bi* on temperature profiles is more pronounced as $Bi \geq 0.03$. The equilibrium time, as defined earlier, for a full pulse varies with changing B_i.

The equilibrium time corresponding to low β', $\frac{\beta'}{\gamma'}$ and high *Bi* is small. This indicates that early rise of the pulse intensity results in small equilibrium time. Moreover, high value of *Bi* reduces the equilibrium time for half and full pulses. In this case, convection cooling of the surface suppresses the internal energy gain of the substrate. Consequently, the domination of the internal energy gain in the energy transport process ceases earlier during the heating pulse.

2.4.2 Repetitive Pulse Heating Case and Convection Condition Resembling Assisting Gas at the Surface

In order to obtain a constant temperature heating condition at the surface, the laser repetitive pulses with constructively decaying peak intensities are considered. To account for the cooling of the surface, the Biot number (Bi) is varied in between 0.2 and 2×10^{-4}, the pulse parameters $\left(\frac{\beta_0}{\gamma_0}\right)$ is varied in between 2.5 and 6, and the ratios of peak pulse intensity of consecutive pulses are $\frac{I_{02}}{I_{01}} = 0.25$ and $\frac{I_{03}}{I_{01}} = 0.0625$, respectively. Table 2.2 gives the data used in the solutions of temperature field. The second pulse of the consecutive pulses starts once the first pulse ends or about to end. The overlapping of the consecutive pulses is minimal. This is, however, more pronounced when $\frac{\beta_0}{\gamma_0}$ reduces as seen from Fig. 2.7, in which the power intensity distribution with time is shown. It should be noted that the energy content of the overlapping portion of the repetitive pulses is only a fraction of the respective repetitive pulses, which overlap.

Figure 2.8 shows the temporal variation of the non-dimensional surface temperature $\left(\frac{T}{I_0/k\delta}\right)$ for three consecutive pulses with $\frac{\beta_0}{\gamma_0} = 4$, and *Bi* as variable while Fig. 2.9 shows the time derivative of temporal variation of the non-dimensional surface temperature for the same condition of Fig. 2.8. The peak temperature increases as *Bi* reduces. This is because the rate of convective cooling of the surface, which reduces as *Bi* reduces, i.e. the heat transfer coefficient reduces. The effect of *Bi* on temperature profile is less pronounced as *Bi* reduces to 10^{-3} and beyond. In this case, the convective cooling of the surface does not substantiate and the internal energy gain dominates over the convective cooling of the surface. As *Bi* increases, the convective cooling of the surface becomes as important as the internal energy gain of the substrate material, which in turn reduces the peak temperature at the surface. When comparing the trend of the temperature profiles

Table 2.2 Data used in the solutions of the repetitive pulse heating

Bi	$\delta \times 10^7$ (1/m)	$\alpha \times 10^{-4}$ (m^2/s)	c_p (J/ kgK)	ρ (kg/ m^3)	k (W/ mK)	β_0 (1/s)	γ_0 (1/s)	I_o (W/m^2)	τ_0
0.2–2 $\times 10^{-4}$	6.16	0.227	460	7,880	80.3	0.3	0.05–0.125	4.5×10^{12}	40

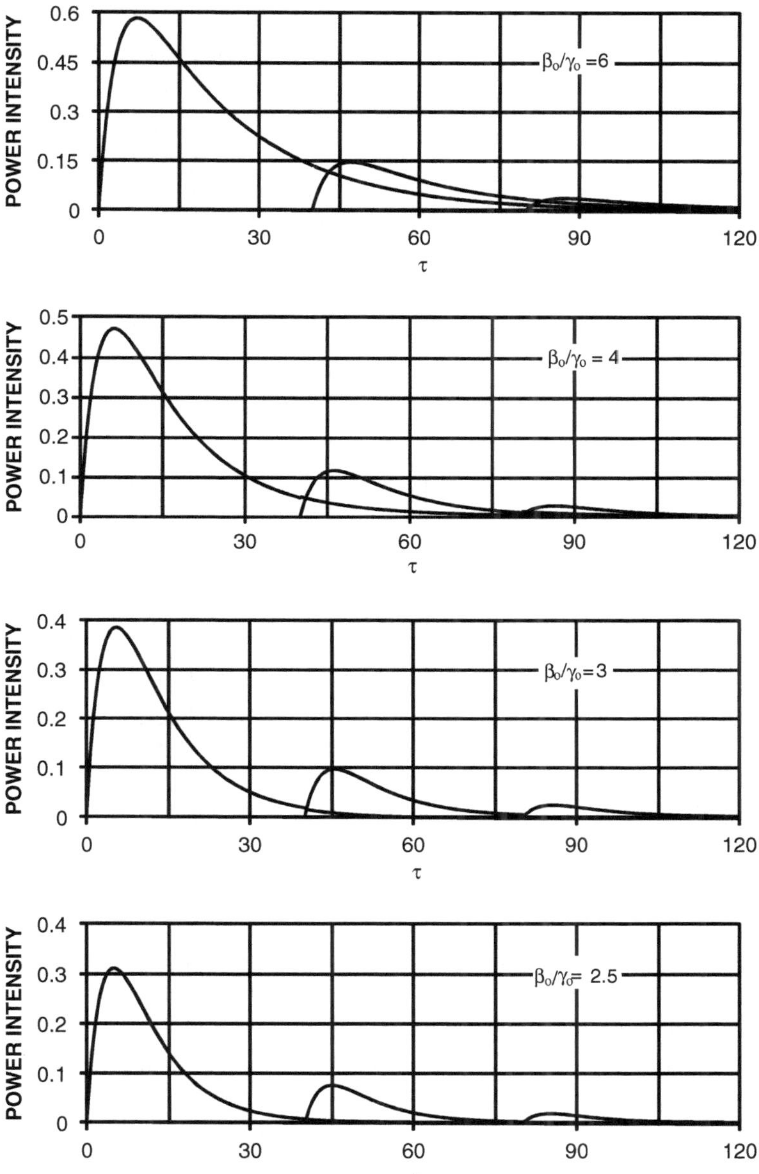

Fig. 2.7 Temporal variation of dimensionless intensity as β_0/γ_0 variable

with time due to consecutive pulses and different *Bi*, the surface temperature shows an increasing trend with time as *Bi* reduces. The trend becomes almost steady with some fluctuations across the peak temperatures for $Bi = 2\times10^{-2}$. The

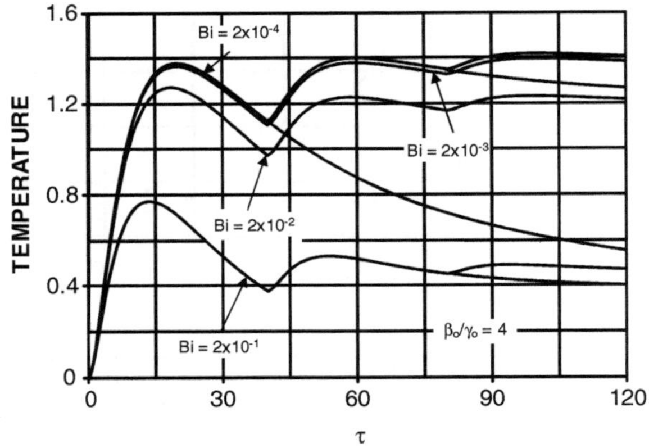

Fig. 2.8 Temporal variation of dimensionless surface temperature for $\beta_0/\gamma_0 = 4$ and Bi variable

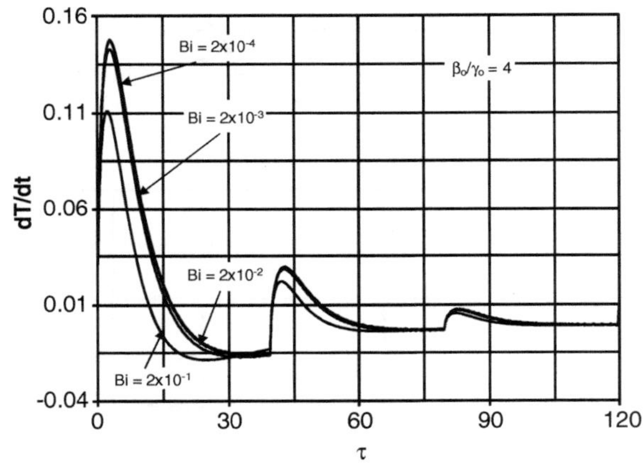

Fig. 2.9 Temporal variation of time derivative of dimensionless surface temperature for $\beta_0/\gamma_0 = 4$ and Bi variable

further reduction in Bi results in decreasing trend with time in surface temperature profile. Consequently, the convective cooling does not lower the peak value of the surface temperature only, but effects the temporal behavior of the surface temperature resulted from the consecutive pulses. In the case of $\frac{\beta_0}{\gamma_0} = 6$ Fig. 2.10, the surface temperature due to consecutive pulses shows increasing trends with time for $Bi \leq 10^{-2}$, and the trend becomes almost steady with time for $Bi = 0.2$. Consequently, increasing $\frac{\beta_0}{\gamma_0}$ results in increasing peak temperature and the surface temperature profile due consecutive pulses shows an increasing trend with time. In the case of high Bi $(Bi = 0.2)$, the surface temperature attain almost steady trend with time.

Figures 2.11, 2.12, and 2.13 show the first peak temperature difference (ΔT_{p1}), second peak temperature difference (ΔT_{p2}), the maximum temperature difference in between the first and second peak temperatures $(\Delta T_{\max1})$, and the maximum temperature difference in between second and third peak temperatures $(\Delta T_{\max2})$

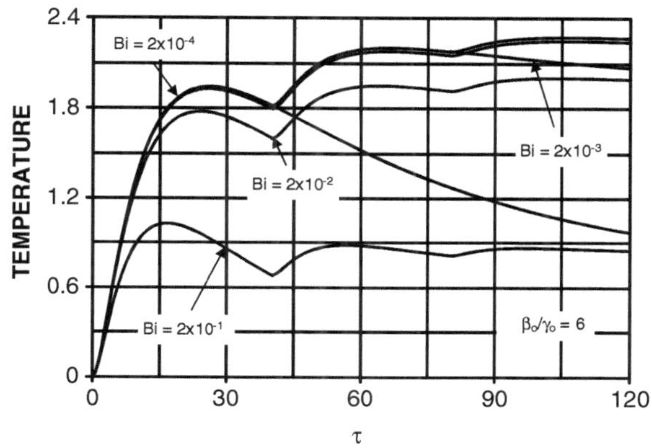

Fig. 2.10 Temporal variation of dimensionless surface temperature for $\beta_0/\gamma_0 = 6$ and Bi variable

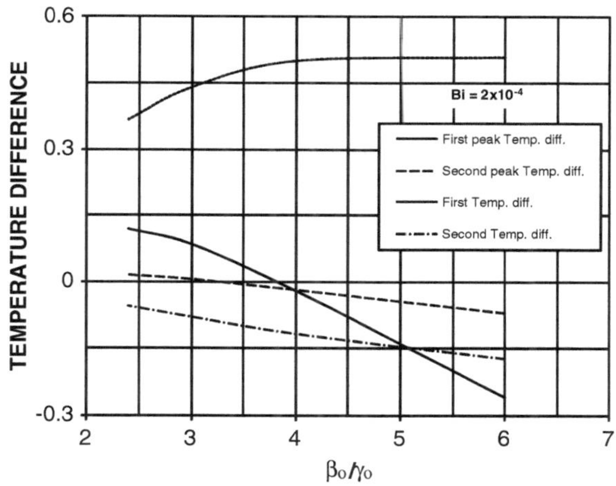

Fig. 2.11 Dimensionless surface temperature difference with β_0/γ_0 and $Bi = 2 \times 10^{-4}$

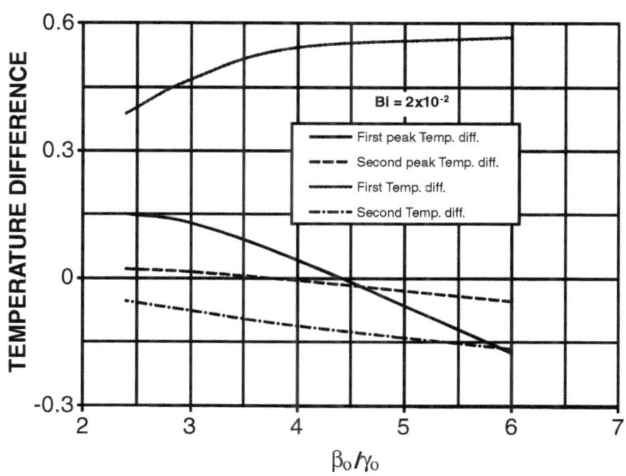

Fig. 2.12 Dimensionless surface temperature difference with β_0/γ_0 and $Bi = 2 \times 10^{-2}$

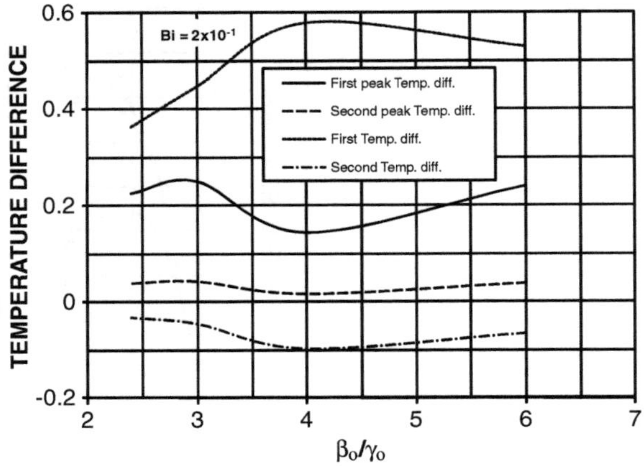

Fig. 2.13 Dimensionless surface temperature difference with β_0/γ_0 and $Bi = 2 \times 10^{-1}$

with $\frac{\beta}{\gamma}$ as Bi variable. ΔT_{p1} is positive for $\frac{\beta_0}{\gamma_0} \leq 4$ and this reverses for $\frac{\beta_0}{\gamma_0} > 4$. This is also true for ΔT_{p2} as $Bi \geq 10^{-3}$. This indicates that the first peak temperature is always greater than the second peak temperature for $\frac{\beta_0}{\gamma_0} < 4$, i.e. the temperature profile shows an increasing trend with time. However, ΔT_{p1} becomes zero for $\frac{\beta_0}{\gamma_0} = 4.5$ for $Bi = 2 \times 10^{-2}$ and $Bi \leq 2 \times 10^{-3}$. In the case of ΔT_{p2}, it becomes zero at $\frac{\beta_0}{\gamma_0} = 3.25$ for $Bi \leq 2 \times 10^{-3}$. Therefore, as Bi increases $\frac{\beta_0}{\gamma_0}$ also increases for the zero value of ΔT_{p1} and ΔT_{p2}. In the case of $Bi = 0.2$, first peak temperature difference is always positive for $\frac{\beta_0}{\gamma_0}$ and this is also true for ΔT_{p2}. Consequently, the surface temperature has a decreasing trend with time. $\Delta T_{\max 1}$ attains large values as Bi reduces. Moreover, $\Delta T_{\max 1}$ increases with increasing $\frac{\beta_0}{\gamma_0}$. As $\frac{\beta_0}{\gamma_0}$ increases, the time corresponding to the peak temperature moves towards the pulse beginning as seen from Fig. 2.8. Consequently, high peak intensity differences results in large peak temperature and maximum temperature differences in the surface temperature profiles. ΔT_{p2} is relatively smaller than ΔT_{p1}. In general, a steady temperature at the surface is unlikely to occur, since $\Delta T_{\max 1}$ and ΔT_{p1} attain high values, which in turn results in large fluctuation in the temperature profile. However, the magnitude of fluctuation becomes less as $\frac{\beta_0}{\gamma_0}$ and Bi reduce.

2.4.3 Effect of Duty Cycle on Heating and Convection Condition Resembling Assisting Gas at the Surface

Table 2.3 gives the data used in the numerical simulations. Figure 2.14 shows temporal variation of surface temperature for different duty cycles of laser pulses. The rise of surface temperature is high in the early heating period for all duty

Table 2.3 Data used for the numerical predictions for the effect of the duty cycle on temperature

$\delta \times 10^7$ (1/m)	c_p (J/kgK)	ρ (kg/m^3)	k (W/mK)	β (1/s)	h (W/m^{2}K)	a (1/m)	γ (1/s)	I_O (W/m^2)
6.16	330	7,836	52	0.1	10^4	0.0707	0.05	4×10^{10}

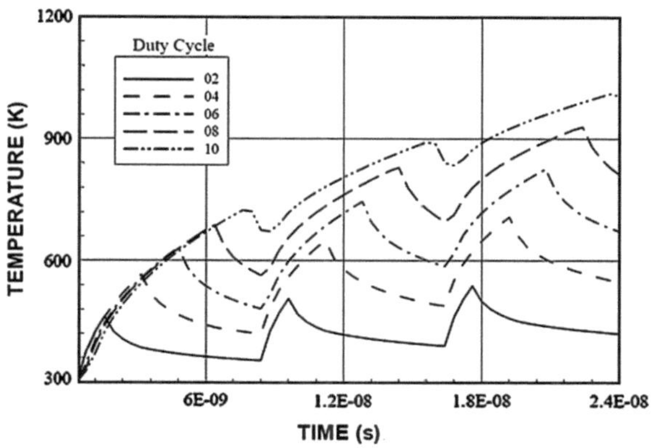

Fig. 2.14 Temporal variation of surface temperature at symmetry axis for the first three pulses

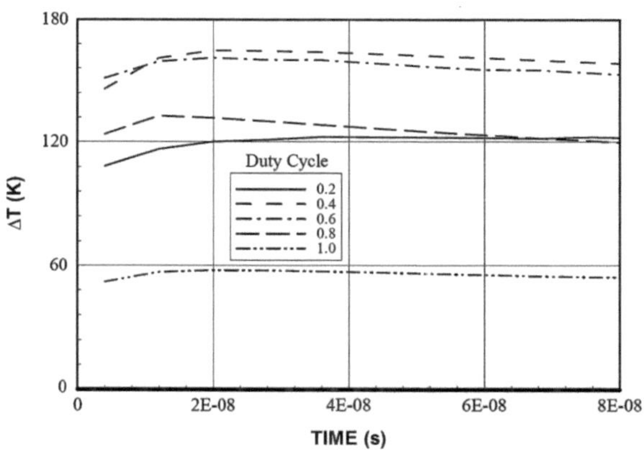

Fig. 2.15 Temporal variation of surface temperature difference (ΔT) at the symmetry axis for the first ten pulses and different duty cycles

cycles and as the pulse progresses towards its ending, the rate of rise becomes steady. The high rate of temperature rise is associated with the internal energy gain and temperature gradient in the surface region of the substrate material. In this case, energy absorbed by the substrate material is high in the surface region due to Lambert's Law, which in turn, increases the internal energy gain substantially in this region. In the early heating period, temperature gradient is low, and diffusional energy transfer, due to temperature gradient, from the surface region to solid bulk becomes less. This enhances the temperature rise. However, as the heating progresses, temperature gradient increases accelerating the diffusional energy transfer from surface region to the solid bulk. Although, increasing duty cycle increases the laser energy in the pulse, enhancing the magnitude of temperature, the rate of

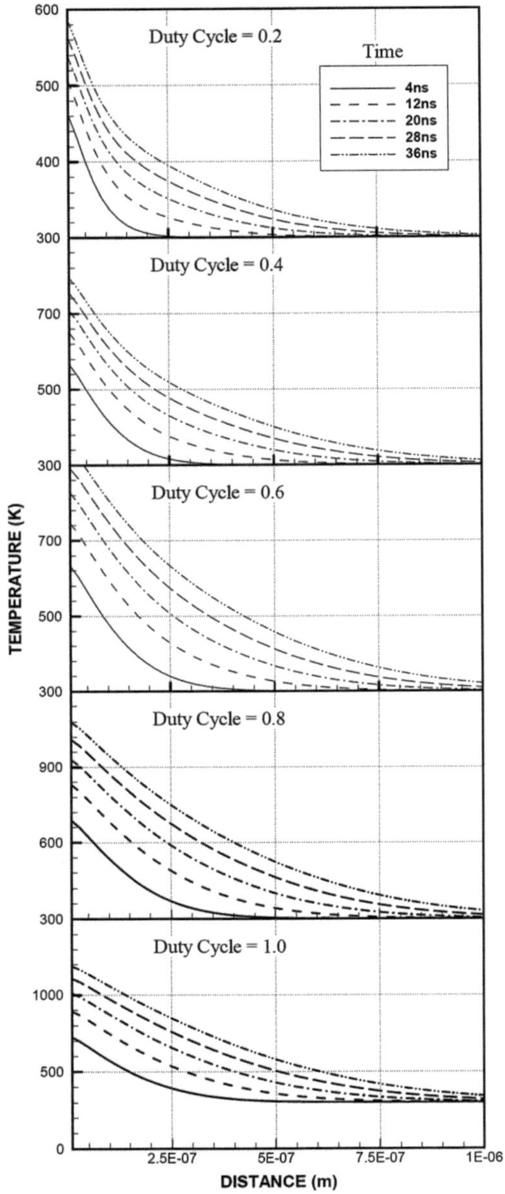

Fig. 2.16 Temperature variation along the symmetry axis inside the substrate material for different heating periods and duty cycles

temperature rise in both early and late heating periods changes. This indicates that increasing duty cycle enhances temperature rise in a short duration at the substrate surface.

Figure 2.15 shows temporal variation of temperature difference at the surface for different duty cycles. Temperature difference is calculated as the difference between the maximum temperature in a pulse and the temperature at the corresponding pulse ending. Temperature difference remains almost constant with time despite the fact that maximum temperature at the surface increases. This indicates that the thermal response of the surface to the heating pulse becomes the same with a given duty cycle of the consecutive pulses. Moreover, as the duty cycle increases, the magnitude of temperature also increases. Once the duty cycle is kept constant for the pulses, the temperature difference remains almost constant with time. The high magnitude of temperature difference results in cyclic thermal loading of the surface. Consequently, thermal integration (steady rise of surface temperature during repetitive pulses) at the surface replaces with thermal loading of the surface. This situation results in thermal fatigue of the surface. Therefore, increasing duty cycle enables surface temperature to rise at high rates; but, it causes cyclic thermal loading at the surface due to magnified temperature difference.

Figure 2.16 shows temperature distribution inside the substrate material for different duty cycles and heating periods when temperature is maximum at the surface. Temperature profiles are plotted when surface temperature is maximum for a given time period. The temperature gradient attains large values, particularly in the surface region for small duty cycles. The temperature gradient attains minimum at some depths below the surface and as the heating duration progresses, the location of minimum temperature gradient moves away from the surface. The energy balance is attained at the point of minimum temperature gradient such that internal energy gain from the irradiated field balances the diffusional energy loss from the surface region to the solid bulk. Consequently, the diffusional energy transfer dominates the energy transfer process in the region beyond the location of minimum temperature gradient. Moreover, increasing duty cycle does not alter the location of minimum temperature gradient considerably in the substrate material. This indicates that the duty cycle change the magnitude of temperature gradient; however, the location of maximum temperature gradient remains almost the same in the substrate material for different duty cycles. Consequently, amount of energy transport in the surface region enhances with increasing duty cycle, but energy transport by diffusion becomes important at the particularly location inside the substrate material for all duty cycles.

References

1. Kalyon, M., Yilbas, B.S.: Analytical solution for laser evaporative heating process: time exponentially decaying pulse case. J. Phys. Part D: Applied Physics **34**, 3303–3311 (2001)
2. Kalyon, M., Yilbas, B.S.: Repetitive laser pulse heating analysis: pulse parameter variation effects on closed form solution. Appl. Surf. Sci. **252**, 2242–2250 (2006)
3. Yilbas, B.S., Shuja, S.Z.: Laser short-pulse heating of surfaces. J. Phys. D Appl. Phys. **32**, 1947–1954 (1999)
4. Blackwell, F.J.: Temperature profile in semi-infinite body with exponential source and convective boundary conditions', ASME. J. Heat Transfer **112**, 567–571 (1990)

5. Shuja, S.Z., Yilbas, B.S., Shazli, S.Z.: Laser repetitive pulse heating influence of pulse duty. Heat Mass Transf. **43**, 949–955 (2007)
6. Yilbas, B.S.: Analytical solution for time unsteady laser pulse heating of semi-infinite solid. Int. J. Mechanical Sciences **39**(6), 671–682 (1997)
7. Yilbas, B.S., Sami, M.: 3-Dimensional laser heating including evaporation—a kinetic theory approach. Int. J. Heat and Mass Transf. **41/13**, 1969–1981 (1998)

Chapter 3
Laser Melting of Solid Surfaces

Abstract Laser processing, in general, is involved with non-conduction heating of substrate materials including the conduction heating, melting, and evaporation. In order to comprehend the physical processes taking place during laser processing, experimenting and/or modeling of the process is necessary. However, experimental tests are costly due to short processing time and high temperature involvement and model studies offer insight into the physical processes, which can be used for process optimization. In this chapter, Laser non-conduction limited heating of solid surfaces is introduced and analytical and numerical approaches for temperature field in the laser induced melted region is presented.

3.1 Introduction

Laser processing, in general, is involved with non-conduction heating of substrate materials including the conduction heating, melting, and evaporation. In order to comprehend the physical processes taking place during laser processing, experimenting and/or modeling of the process is necessary. However, experimental tests are costly due to short processing time and high temperature involvement and model studies offer insight into the physical processes, which can be used for process optimization. When modeling the laser non-conduction heating process, either the volumetric or the surface heat sources are considered. In actual situation, laser beam intensity is absorbed within the surface skin of the substrate material and the absorbed energy is re-distributed inside the substrate material according to the Beer-Lambert's Law. The depth of absorption is shallow for metals, which is in the order of 10^{-8} m. Since the absorption results in exponential decay of laser intensity within the absorption depth, volumetric heat source consideration in the model study becomes necessary to capture the actual heating situation.

In most of the industrial applications, laser beam scans the surface with a constant velocity and the solid heating and phase change in the irradiated region extends

B. S. Yilbas and S. Z. Shuja, *Laser Surface Processing and Model Studies*, Materials Forming, Machining and Tribology, DOI: 10.1007/978-3-642-36629-1_3,

along the direction of the laser movement. The progressing of the phase change over the thickness results in cutting or welding of the workpiece. However, the laser surface treatment process requires controlled melting in the surface vicinity. In this chapter, laser non-conduction limited heating and analytical treatment of melting process is presented. In addition, numerical analysis pertinent to non-conduction limited heating for a stationary and a moving laser sources is included.

3.2 Analytical Treatment of Laser Melting Process

One dimensional heating situation can be incorporated for a stationary pulsed layer non-conduction limited heating of a solid substrate. This is because of the fact that the radial temperature gradient is considerably smaller than that corresponding to the axial temperature gradient in the irradiated region. In addition, the depth of laser irradiated region is considerably smaller than the thickness of the substrate material; therefore, the workpiece can be considered as a semi-infinite body. In order to present the governing equations of heat transfer in a simple form and obtained a closed form solution for the temperature rise, some useful assumptions can be made. These include liquid ejection and nucleation in the liquid phase are neglected, thermal properties of the substrate are considered to be constant, and the radiative cooling of the surface is ignored. It should be noted that an experiment is carry out to verify the ablation process that is governed purely by the vapor ejection during the nano-second laser heating pulse, i.e., no significant liquid ejection took place during the nanosecond heating process [1]. The absorption of the incident laser beam in the vapor phase is also omitted.

3.2.1 The Closed Form Solution

The mathematical analysis pertinent to non-conduction limited heating of laser irradiated surface is presented in line with the previous study [2]. Since the heating analysis is carried out for nanosecond laser pulses, which is longer than the thermalization time of the solids, the Fourier theory of heating is still applicable when modeling the heating process. Therefore, the heat transfer equation for a laser step input heating pulse due to semi-infinite solid can be written as:

$$k\frac{\partial^2 T(x,t)}{\partial x^2} + \rho CpV\frac{\partial T(x,t)}{\partial x} + I_{surf}\delta\exp(-\delta x) = \rho c_p\frac{\partial T(x,t)}{\partial t} \quad (3.1)$$

where k is the thermal conductivity, ρ is the density, Cp is the specific heat, $I_{surf} = I_o(1 - r_f)$, which is laser power intensity available at the surface, I_o is the laser power intensity, and r_f is the surface reflectivity, V is the liquid/solid interface velocity (recession velocity) of the surface, which can be determined from either (1) a constant temperature melting at the surface, or (2) a steady heating situation resulting in the energy balance at the surface [3].

Initially the substrate material is considered at a uniform temperature with zero magnitude. Therefore, the initial condition is:

$$\text{Initial time} \quad t = 0 \quad \Rightarrow \quad T(x, 0) = 0$$

The convective boundary condition is considered at the surface of the substrate material due to melting at the surface and at a depth infinitely below the surface, temperature is assumed to reduce to the initial temperature. Therefore, the boundary conditions are:

$$\begin{array}{lll} x \text{ at infinity} & x = \infty & \Rightarrow \quad T(\infty, t) = 0 \\ \text{At the surface} & x = 0 & \Rightarrow \quad k\frac{\partial T(r,0,t)}{\partial x} = \rho V L_m \end{array}$$

where L_m is the latent heat of melting.

Equation 3.1 is in a non-linear form, since the velocity V is changing with time. Whilst a complete analytic unsteady solution to the evaporating boundary problem appears unlikely, a worthwhile solution to the quasi-steady problem is feasible. Laplace transformation of Eq. 3.1 with respect to time allows the resultant second order differential equation to be solved in the transformed plane. The mathematical arrangements of the derivation of closed form solution are referred to [2]. The resulting closed form solution is:

$$\begin{aligned} T(x, t) = & \frac{I_{surf}\delta\sqrt{\alpha}}{2\rho C_p(\alpha\delta - V)} \left\{ \begin{array}{c} 4\sqrt{t}\,\mathrm{ierfc}\left(\frac{x}{2\sqrt{\alpha t}} + b\sqrt{t}\right) \\ + \frac{3b^2 + c^2}{2b(b^2 - c^2)}\,\mathrm{erfc}\left(\frac{x}{2\sqrt{\alpha t}} + b\sqrt{t}\right) \\ + \frac{1}{2b}\exp\left(-\frac{2bx}{\alpha}\right)\,\mathrm{erfc}\left(\frac{x}{2\sqrt{\alpha t}} - b\sqrt{t}\right) \\ + \frac{1}{(b-c)}\exp[-[\delta x + (b^2 - c^2)]t]\,\mathrm{erfc}\left[-\left(\frac{x}{2\sqrt{\alpha t}} + c\sqrt{t}\right)\right] \\ - \frac{1}{(b+c)}\exp\left[\frac{x}{\sqrt{\alpha}}(b + c) + (b^2 - c^2)t\right]\,\mathrm{erfc}\left(\frac{x}{2\sqrt{\alpha t}} - c\sqrt{t}\right) \\ - \frac{2}{(b-c)}\exp(-\delta x) \end{array} \right\} \\ & - \frac{\rho V L_m}{4bk} \left\{ \begin{array}{c} 4b\sqrt{\alpha t}\,\mathrm{ierfc}\left(\frac{x}{2\sqrt{\alpha t}} + b\sqrt{t}\right) \\ -\sqrt{\alpha}\,\mathrm{erfc}\left(\frac{x}{2\sqrt{\alpha t}} + b\sqrt{t}\right) + \sqrt{\alpha}\exp\left(-\frac{-2bx}{\sqrt{\alpha}}\right)\,\mathrm{erfc}\left(\frac{x}{2\sqrt{\alpha t}} - b\sqrt{t}\right) \end{array} \right\} \end{aligned} \tag{3.2}$$

where $b = \frac{V}{2\sqrt{\alpha}}$ and $c = b - \delta\sqrt{\alpha}$. erfc $\mathrm{erfc}(\chi)$ is the complementary error function, which is:

$$\mathrm{erfc}(\chi) = 1 - \mathrm{erf}(\chi) \tag{3.3}$$

$\mathrm{ierfc}(\chi)$ is the integral of complementary error function, which is:

$$ierfc(\chi) = \chi erfc(\chi) - \frac{1}{\sqrt{\pi}} \exp(-\chi^2) \tag{3.4}$$

The quasi-steady solution for the temperature (Eq. 3.2) can be used to form the basis for a more accurate solution. In this case, two approaches are introduced. In the first approach, surface temperature is set to melting temperature of the substrate material and recession velocity is determined from the boundary conditions at the surface. In the second case, recession velocity is calculated from the energy equilibrium consideration at the surface (steady heating situation; in which case, the laser power available is absorbed at the surface) [3], which is:

$$V = \frac{I_{surf}}{\rho[CpT_s + L_m]} \tag{3.5}$$

where T_s is the surface temperature. Consequently, Eqs. 3.2 and 3.5 are solved simultaneously to obtain the surface temperature and recession velocity for various laser power intensities.

3.2.2 *Influence of Assisting Gas on the Melt Layer Thickness*

The analysis associated with the influence of assisting gas including cooling and exothermic reaction contribution on the melt formation is examined in line with the previous study [4]. During the melting process, the assisting gas forms a boundary layer flow over the liquid surface (molten metal surface). In the case of inert assisting gas, the heat transfer from liquid surface to boundary layer flow occurs at assisting gas–liquid interface, since the assisting gas is at room temperature, and heat transfer from liquid surface to solid substrate occurs at liquid–solid interface. Moreover, the contribution of oxidation reactions in the melting section can be neglected for laser inert gas assisted processing.

The energy balance associated with the melting process can be simplified through investigating the melting process by lumped parameter technique. Therefore, it is considered that the melt layer generated flows steadily in the direction of the assisting gas due to the drag force developed at the assisting gas–liquid interface and the bulk properties of the substrate material are assumed to be independent of temperature. Figure 3.1 shows a schematic view of the melt flow.

Consider the melt layer generated at the solid surface during the steady laser heating process. Assume a small fraction (β_{ev}) of molten metal evaporates from the surface at the assisting gas–melt interface during the laser heating process. The rate of energy required $(\dot{E}_{req})$ to generate a melt flow rate $(\dot{m}_L)$ at the surface of the solid substrate can be written as:

$$\dot{E}_{req} = \dot{m}_L[Cp_s(T_m - T_i) + L_m + \beta_{ev}L_{ev} + 1.65Cp_m(T_{ev} - T_m)] \tag{3.6}$$

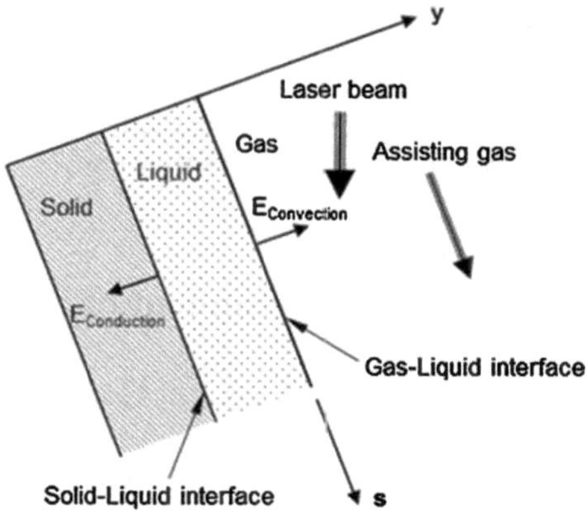

Fig. 3.1 Schematic view of the melt section

where T_i and T_m are initial and the melting temperatures of the solid substrate. L_m and L_{ev} are the latent heating of melting and evaporation of the substrate material. Cp_m and Cp_s are the specific heat capacity of the liquid and solid phases of the substrate. It should be noted that since the liquid layer flows along the surface due to a shear force across the gas–liquid interface, it stores extra energy. Therefore, the term $0.65Cp_m(T_{ev} - T_m)$ represents the extra rate of energy due to melt layer flow at the surface as approximated in the previous study [5]. The formulation of the liquid layer thickness is provided in the previous formulation and the resulting equation is presented below, i.e.:

$$\frac{\delta_L}{s} = \sqrt{\left[\mu_L \frac{2}{\delta_L U_e}\left(C_1 - \frac{5}{4U_e}\frac{C_3}{C_2}\right)\frac{2}{C_f \rho_g U_e}\frac{1}{s}\right]} \tag{3.7}$$

where δ_L is the liquid layer distance, s is the distance along the laser scanning direction, μ_L is the viscosity of the molten flow, U_e is the assisting gas velocity above the surface, C_f is the friction factor, ρ_g is the assisting gas density and C_1, C_2, and C_3 are:

$$C_1 = \frac{1}{\rho_L} \frac{\frac{P_o}{A} + \dot{q}_C - h(T_m - T_{oi})}{[Cp_s(T_m - T_i) + L_m + \beta_{ev} L_{ev} + 1.65Cp_m(T_{ev} - T_m)]} \tag{3.8}$$

and

$$C_2 = \rho_L[Cp_s(T_m - T_i) + L_m + \beta_{ev} L_{ev} + 1.65Cp_m(T_{ev} - T_m)] \tag{3.9}$$

and

$$C_3 = \rho_g U_e C_H \left[Cp_g(T_{ev} - T_{oe}) + \frac{1}{2}\left(U_e^2 - U_{LS}^2\right)\right] \tag{3.10}$$

Here, Cp_s is the specific heat of the solid substrate, P_o is the laser power at the workpiece surface, A is the cross-sectional area of the melt layer, $\dot{q}_C$ is the rate of heat conduction, T_m is the melting temperature, β_{ev} is the fraction of evaporation at the surface, T_{oi} is the ambient temperature, C_H is the thermal conductance, Cp_g is the specific heat of the assisting gas, U_{LS} is the liquid layer velocity.

3.3 Numerical Treatment of Surface Melting

The depth of the melt pool in laser surface treatment application, in general, is limited within tens of micrometers and high temperature involved makes it difficult to monitor the melt pool formation with high accuracy during the melting. Although analytical studies provide information on the functional relation between the dependent variables and the process variables, analytical solutions are limited due to the simplifications and assumptions introduced in the analysis. Numerical treatment of the heating process reduces the assumptions and the simplifications in analysis; thereby, it provides useful information on the physical processes taking place during the laser heating in relation to melting.

3.3.1 Stationary Heating Source

In the melt layer a flow field is developed due to the Marangoni effect. The convection current developed in the melt pool influences the rate of heat transfer while modifying the melt pool size and the mushy zone thickness. This, in turn, influences the laser melting process and the resulting surface quality. Therefore, the model study should include the Marangoni flow in the melt pool unlike the analytical solution presented in Eq. 3.2.

3.3.1.1 Heating Analysis

The mathematical arrangements of the heating process can be formulated incorporating the axisymmetric heating of a semi-infinite body with volumetric heat source resembling the laser beam irradiation in line with the previous study [6]. Since the laser melting process is involved with heating and phase change only, the enthalpy equation can be written as:

$$\frac{\partial(\rho H)}{\partial t} + \nabla \cdot (\rho \bar{v} H) = \nabla \cdot (k \nabla T) + S_o \tag{3.11}$$

and S_o is the volumetric heat source term and it is:

$$S_o = I_o \delta (1 - r_f) \exp(-\delta x) \exp\left(-\left(\frac{r}{a} + \beta\right)^2\right) \tag{3.12}$$

where H is the enthalpy, ρ is the density, k is the thermal conductivity, $\bar{v}$ is the velocity vector in the melt pool I_o is laser peak intensity, δ is the absorption coefficient, a is the Gaussian parameter, r_f is the surface reflectivity. It is assumed that the laser pulse intensity can be considered as a step input intensity, since the rise and decay of the laser pulse is negligibly small as being the case for some of the practical lasers [7].

The laser beam axis is the x-axis and the solution domain is shown in Fig. 3.2. However, in the solid phase heating the velocity vector is set to zero. Equation 3.11 is solved numerically with the appropriate boundary conditions to predict the temperature field in the substrate material.

Initially, it is assumed that the workpiece is at uniform temperature:

$$\text{Initial time} \quad t = 0 \quad \Rightarrow \quad T(r, x, 0) = T_0(\text{specified})$$

The boundary conditions for the temperature field are:

$$x \text{ at infinity} \quad x = \infty \quad \Rightarrow \quad T(r, \infty, t) = T_0(\text{specified})$$

$$r \text{ at infinity} \quad r = \infty \quad \Rightarrow \quad T(\infty, x, t) = T_0(\text{specified})$$

$$\text{At symmetry axis} \quad r = 0 \quad \Rightarrow \quad \frac{\partial T(0,x,t)}{\partial r} = 0$$

$$\text{At the surface} \quad x = 0 \quad \Rightarrow \quad k\frac{\partial T(r,0,t)}{\partial x} = h_t(T_s - T_\infty) + \sigma_t(T_s - T_\infty)$$

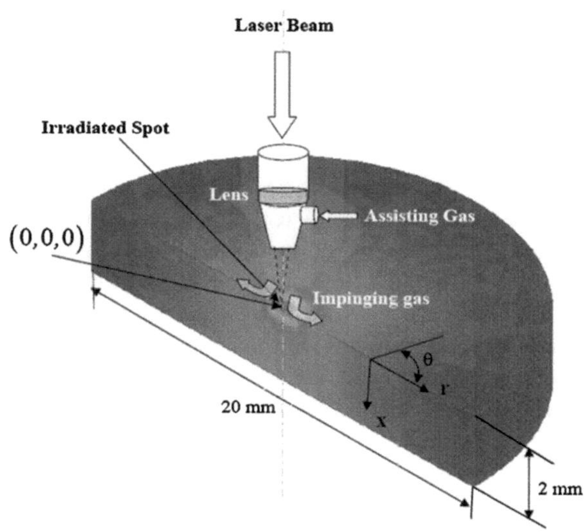

Fig. 3.2 A schematic view of the laser heating and stationary heat source consideration

where h_t is the heat transfer coefficient at the free surface, $\sigma = 5.67 \times 10^{-8}\,\mathrm{W/m^2K^4}$ is the Stefan–Boltzmann constant. The heat transfer coefficient predicted earlier [8] is used in the present simulations $\sigma = 5.67 \times 10^{-8}\,\mathrm{W/m^2K^4}$.

To analyze the phase change problem, the enthalpy-porosity technique is used. In this case, the melt interface is tracked explicitly after defining a quantity called the liquid fraction (β), which indicates the fraction of the cell volume that is in liquid form. Based on the enthalpy balance, the liquid fraction is computed. The mushy zone is a region in which the liquid fraction lies between 0 and 1. The mushy zone is modeled as a "pseudo" porous medium in which the porosity decreases from 1 to 0 as the material solidifies. When the material has fully solidified in a cell, the porosity becomes zero and hence the velocities also drop to zero. The enthalpy of the material is computed as the sum of the sensible enthalpy, h, and the latent heat, ΔH:

$$H = h + \Delta H \tag{3.13}$$

where

$$h = h_{ref} + \int_{T_{ref}}^{T} C_p dT \tag{3.14}$$

and h_{ref} is the reference enthalpy, T_{ref} is the reference temperature, and Cp is the specific heat capacity. The liquid fraction, β_l, can be defined as:

$$\begin{array}{ll} \beta_l = 0 & T < T_{solidus} \\ \beta_l = 1 & T > T_{liquidus} \\ \beta_l = \frac{T - T_{solidus}}{T_{liquidus} - T_{solidus}} & T_{solidus} < T < T_{liquidus} \end{array} \tag{3.15}$$

Equation 3.15 is referred to as the lever rule [9].

The latent heat content can be written in terms of the latent heat of melting of the material (L):

$$\Delta H = \beta_l L_m \tag{3.16}$$

The latent heat content can vary between zero (for a solid) and L_m (for a liquid).

The enthalpy-porosity technique treats the mushy region (partially solidified region) as a porous medium. The porosity in each cell is set equal to the liquid fraction in that cell. In fully solidified regions, the porosity equals to zero, which extinguishes the velocities in these regions. The flow in the mushy zone is governed by the Darcy law. In this case, as the porosity in the mushy zone decreases, the permeability and the velocity also decrease, i.e., when the mushy zone becomes completely solid at the interface of the mushy zone-solid phase, velocity reduces to zero. This behavior can be accounted for by defining the momentum sink as $S_z = A \cdot \bar{v}$ where A is obtained from Carman-Koseny equation [10]; in which case, it is shown that $A = \frac{(1-\beta_l)^2}{(\beta_l^3+\varepsilon)} A_{mush}$ [10]. Therefore, the momentum sink due to the reduced porosity in the mushy zone takes the following form [9, 10]:

$$S_z = \frac{(1-\beta_l)^2}{(\beta_l^3+\varepsilon)} A_{mush}(\bar{v}) \tag{3.17}$$

where β_l is the liquid volume fraction, ε is a small number (0.001) to prevent division by zero, A_{mush} is the mushy zone constant. The mushy zone constant measures the amplitude of the damping; the higher this value, the steeper the transition of the velocity of the material to zero as it solidifies. The liquid velocity can be found from the average velocity is determined from:

$$\vec{v}_{liq} = \frac{\vec{v}}{\beta_l} \tag{3.18}$$

The solution for temperature is essentially iteration between the enthalpy equation (Eq. 3.11) and the liquid fraction equation (Eq. 3.17). Directly using Eq. 3.17 to update the liquid fraction usually results in poor convergence of the energy equation. However, the method suggested by Voller and Prakash [10] is used to update the liquid fraction based on the specific heat.

The continuity and momentum equations in the melt layer are different than that corresponding to the mushy zone. Therefore, the conservation equations need to be incorporated to account for the flow field. This problem is governed by the axi-symmetric Navier–Stokes equation. For laminar flow, the conservation equations are:

$$\begin{aligned}
&\textit{Continuity} && \frac{\partial u}{\partial x} + \frac{1}{r}\frac{\partial (rv)}{\partial r} = 0\\
&r-dir && \frac{\partial v}{\partial t} + u\frac{\partial v}{\partial x} + v\frac{\partial v}{\partial r} = -\frac{1}{\rho}\frac{\partial p}{\partial r} + \frac{\mu}{\rho}\left(\frac{\partial^2 v}{\partial x^2} + \frac{1}{r}\frac{\partial}{\partial r}\left(r\frac{\partial v}{\partial r}\right) - \frac{v}{r^2}\right)\\
&x-dir && \frac{\partial u}{\partial t} + u\frac{\partial u}{\partial x} + v\frac{\partial u}{\partial r} = -\frac{1}{\rho}\frac{\partial p}{\partial x} + \frac{\mu}{\rho}\left(\frac{\partial^2 u}{\partial x^2} + \frac{1}{r}\frac{\partial}{\partial r}\left(r\frac{\partial u}{\partial r}\right)\right) + g\beta_E T + S_z\\
&\textit{Energy} && \rho\frac{\partial (H)}{\partial t} + \nabla\cdot(\rho\,\bar{v}\,H) = \nabla\cdot(k\nabla T) + S_o - S_h\\
& && \text{or } \frac{\partial T}{\partial t} + u\frac{\partial T}{\partial x} + v\frac{\partial T}{\partial r} = \alpha\left(\frac{\partial^2 T}{\partial x^2} + \frac{1}{r}\frac{\partial}{\partial r}\left(r\frac{\partial T}{\partial r}\right)\right) + S_o - S_h
\end{aligned} \tag{3.19}$$

where u and v are the component of the velocity in the radial r and axial x directions, respectively, p is the pressure, T is the temperature, α is the thermal diffusivity of the molten material, β_E is the volumetric thermal expansion coefficient, which is formulated using the Boussinesq approximation, S_z is the momentum sink due to the reduced porosity in the mushy zone (Eq. 3.19), and S_h is a phase related source term convection–diffusion phase change ($S_h = \frac{\partial(\rho\Delta H_s)}{\partial t} + \nabla\cdot(\rho\,\bar{v}\,\Delta H_s)$, where $\Delta H_s = H - \rho C_p T$) [10]. The term $g\beta_E T$ is the buoyancy term, which is used to introduce natural convection in the melt pool. The energy equation is similar to Eq. 3.12 is used for the phase change and the liquid phase during the laser heating process. Moreover, a zero reference temperature has been assumed for the buoyancy force term in the z-direction momentum equation. Boundary conditions for Eq. 3.19 are as follows:

$$\text{Free surface} \quad x = 0 \quad \Rightarrow \quad \mu \frac{\partial v}{\partial x} = -\sigma_T \frac{\partial T}{\partial r}; \quad u = 0; \quad -k \frac{\partial T}{\partial x} = q_{loss}$$

where μ is the dynamic viscosity, σ_T is the surface tension, and k is the thermal conductivity. q_{loss} includes the convective losses from the surface. The heat transfer coefficient is taken as $h_t = 10^4$ W/m^{2}K at the surface [8].

$$\text{Axis of symmetry} \quad r = 0 \quad \Rightarrow \quad v = 0; \quad \frac{\partial u}{\partial r} = 0; \quad \frac{\partial T}{\partial r} = 0$$

$$\text{Far field (solid)} \qquad T = T_0$$

where T_o is the ambient temperature. The equation $\mu \frac{\partial v}{\partial x} = -\sigma_T \frac{\partial T}{\partial r}$ states the balance between the surface tension force and the viscous force on the free surface.

3.3.1.2 Numerical Analysis

To discretize the governing equations, a control volume approach is introduced. The details of the numerical scheme are given in [11]. The calculation domain is divided into grids and a grid independence test is performed for different grid sizes and orientation. A non-uniform grid with 90 × 55 mesh points along x and r-axes, respectively, is employed after securing the grid independence. It should be noted that to secure the grid independent solutions, the grid selected is doubled in number along both axes (x and r) and the maximum temperatures obtained due to the grid selected and the doubled sized grid are compared. It is found that the change in the maximum temperatures is 0.7 %, which is negligibly small. Therefore, the current grid is used in the simulations. The finer grids are located near the irradiated spot center in the vicinity of the surface and grids become coarse as the distance increases towards the bulk of substrate material. The central difference scheme is adopted for the diffusion terms. The convergence criterion for the residuals is set as $\left|\psi^k - \psi^{k-1}\right| \le 10^{-6}$ to terminate the simulations. The laser power is kept the same as the experimental value in the simulations, i.e., Gaussian parameter $a = \frac{2}{3}R$ (R is a laser beam radius at the workpiece surface $R = 0.3$ mm) and power intensity $I_0 = 0.9 \times 10^9$ W/m^2. The simulations are carried out for a single laser pulse, which ends after 1 ms.

3.3.2 *Influence of Pulse Profile on Temperature Field*

In actual situation, laser beam intensity is absorbed within the surface skin of the substrate material and the energy absorbed is distributed inside the substrate material according to the Beer-Lambert's Law. The depth of absorption is shallow for metals, which is in the order of 10^{-8} m. Since the absorption results in exponential decay of laser intensity within the absorption depth, volumetric heat source consideration in the model study becomes necessary to capture the actual heating situation. Moreover, since the absorption depth is significantly small, the numerical simulation of the heating process requires an excessive number of grid points (large mesh size) in the solution domain due to fine grid spacing in the irradiated region. This requires a computational memory and increases the time required for the converged solution. However, consideration of laser intensity as the surface heat source overcomes this situation and minimizes the computational efforts. However, the accuracy of the predictions becomes in question. Consequently, investigation into the comparison of the influence of the volumetric and surface heat sources, having the same power intensity, on the temperature field becomes necessary. Moreover, the phase change during the heating needs to be included in the model study.

3.3.2.1 Heating Analysis

Two cases are considered for the sources of heating of the substrate material. In the first case, the volumetric heat source is considered and the surface heat source is incorporated in the second case in line with the previous study [12]. The energy equation (Eq. 3.11) is used incorporating the volumetric source. In addition, the volumetric heat source can be arranged to resemble the laser repetitive pulses, i.e.,

$$S_o = I_o \delta \left(1 - r_f\right) \exp(-\delta x) \exp\left(-\left(\frac{r}{a} + \beta\right)^2\right) f(t) \tag{3.20}$$

where I_o is the laser peak power intensity, $\delta$is the absorption coefficient, r_f is the reflectivity, x is the axial distance, β is the laser pulse parameter, a is the Gaussian beam parameter, and $f(t)$ is the temporal distribution of laser pulse intensity, respectively. However, the laser pulse energy is kept constant for all the values of the laser pulse intensity parameters considered in the simulations. β is associated with the spatial distribution of the laser pulse intensity at the workpiece surface; in which case, $\beta = 0$ represents the Gaussian beam profile at the workpiece surface. The time function $f(t)$ representing the step input pulse is:

$$f(t) == \left\{ \begin{array}{ll} 0, & t = 0 \\ 1, & 0 < t \le t_p \\ 0, & t > t_p \end{array} \right\} \tag{3.21}$$

where t_p is the pulse length.

The energy equation incorporating the surface heat source model differs from Eq. (3.11) due to the surface heat source consideration. In this case, enthalpy equation becomes:

$$\frac{\partial}{\partial t}(\rho H) + \nabla \cdot \left(\rho \vec{v} H\right) = \nabla \cdot (k \nabla T). \tag{3.22}$$

Surface heat source, resembling the laser beam, is introduced as a boundary condition at the surface; therefore, the surface heat source is the same as the conduction flux at the surface, i.e.:

The heat flux at the surface

$$x = 0 : k \frac{\partial T(r, 0, t)}{\partial x} = S_o \tag{3.23}$$

and

$$S_o = I_o \delta (1 - r_f) \exp\left(-\left(\frac{r}{a} + \beta\right)^2\right) f(t) \tag{3.24}$$

where $f(t)$ is the same as in Eq. 3.21.

3.3.2.2 Numerical Analysis

Numerical treatment of energy equation is similar to that presented earlier. The solution for temperature is essentially iteration between the energy equation, Eq. (3.11) for volumetric source and Eq. (3.20) for the surface source, and the liquid fraction equation (Eq. 3.17). Directly using Eq. (3.17) to update the liquid fraction usually results in poor convergence of the energy equation; therefore, the method suggested by Voller and Prakash [10] is used to update the liquid fraction based on the specific heat. The calculation domain is divided into grids and a grid independence test is performed for different grid sizes and orientation. A non-uniform grid with 350 × 450 mesh points along x and r axes, respectively, is employed for volumetric heat source model after satisfying the convergence criteria. However, a grid with 350 × 450 mesh points along x and r axes, respectively, is incorporated for volumetric heat source model while 70 × 90 mesh size (along x and r axes, respectively) is used for the surface heat source model. The grid independent results are secured for both grids used in the simulations. The finer grids are located near the irradiated spot center in the vicinity of the surface and grids become courser as the distance increases towards the bulk of substrate material. The central difference scheme is adopted for the diffusion terms. The convergence criterion for the residuals is set as $\left|\psi^k - \psi^{k-1}\right| \leq 10^{-6}$ to terminate the simulations; consequently, the converged results are obtained and presented.

3.3.3 *Influence of Marangoni Flow on Temperature Field*

In laser surface treatment process, in general, an assisting gas is used to protect the surface from the oxidation during melting, i.e., the assisting gas is inert, which in turn prevents the surface from high temperature oxidation reactions. Moreover, the assisting gas also cools the surface during the process. Consequently, investigation into the effect of assisting gas velocity on the melt depth and melt flow velocity during laser heating process is necessary. The mathematical arrangements of the heating problem are similar to those presented in Sect. 3.3.2. However, the flow equations (Eq. 3.19) for the Marangoni effect are omitted in the melt pool for the non-flow field melt pool simulations.

3.3.4 *Moving Heat Source Consideration*

In most of the industrial applications, laser beam scans the surface with a constant velocity and the solid heating and phase change in the irradiated region extends along the direction of the laser movement. However, the laser surface treatment process requires controlled melting in the surface vicinity. Therefore, investigation into laser heating of solid surfaces including the laser scanning speed and phase change in the irradiated material becomes essential.

3.3.4.1 Mathematical Analysis:

The mathematical analysis related to laser melting of the surface due to the moving source is presented in line with the previous study [12]. The heat transfer equation in relation to the laser heating process can be written as:

$$\rho \frac{DE}{Dt} = (\nabla(k\nabla T)) + S_o \tag{3.25}$$

where E is the energy gain by the substrate material, ρ is the density k is the thermal conductivity, and S_o is the volumetric heat source term and it is:

$$S_o = I_o\delta\left(1 - r_f\right)e^{-\delta z}e^{-\left(\frac{x^2+y^2}{a^2}\right)} \tag{3.26}$$

I_o is laser peak intensity, δ is the absorption depth, a is the Gaussian parameter, r_f is the surface reflectivity, and x and y are the axes while the laser beam scans the surface along the x-axis. The laser beam axis is the z-axis (Fig. 3.3).

In the case of a moving heat source along the x-axis with a constant velocity U, energy gain by the substrate material yields:

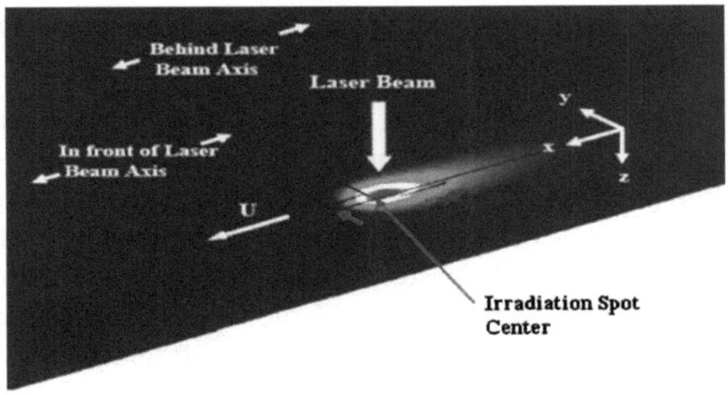

Fig. 3.3 Schematic view of laser heating by a moving laser source and coordinate system

$$\rho \frac{DE}{Dt} = \rho \frac{\partial E}{\partial t} - \rho U \frac{\partial E}{\partial x} \tag{3.27}$$

or

$$\rho \frac{DE}{Dt} = \rho \frac{\partial (CpT)}{\partial t} - \rho U \frac{\partial (CpT)}{\partial x} \tag{3.28}$$

Combining Eqs. (3.25) and (3.28) yields:

$$\rho \frac{\partial (CpT)}{\partial t} = (\nabla (k \nabla T)) + \rho U \frac{\partial (CpT)}{\partial x} + \mathrm{S_o} \tag{3.29}$$

Since the laser scanning speed remains constant, the heating situation can be considered as a steady; in which case the term $\rho \frac{\partial (CpT)}{\partial t}$ in Eq. (3.29) reduces to zero $\left(\rho \frac{\partial (CpT)}{\partial t} = 0\right)$.

Equation (3.29) is solved numerically with the appropriate boundary conditions to predict the temperature field in the substrate material. However, to analyze the phase change problem, the enthalpy-porosity technique is used. In this case, the melt interface is tracked explicitly after defining a quantity called the liquid fraction, which indicates the fraction of the cell volume that is in liquid form. Based on the enthalpy balance, the liquid fraction is computed. The mushy zone is a region in which the liquid fraction lies between 0 and 1. The mushy zone is modeled as a "pseudo" porous medium in which the porosity decreases from 1 to 0 as the material solidifies. When the material has fully solidified in a cell, the porosity becomes zero and hence the velocities also drop to zero. The phase change process uses the enthalpy method as described in Sect. 3.3.1. Therefore, the analysis related to the phase change is referred to Sect. 3.3.1

3.3.4.2 Numerical Analysis:

To discretise the governing equation, a control volume approach is introduced. The calculation domain is divided into grids and a grid independence test is performed for different grid sizes and orientation. A non-uniform grid with 90 × 45 × 55 mesh points along x, y and z axes, respectively, is employed after securing the grid independence. The finer grids are located near the irradiated spot center in the vicinity of the surface and grids become courser as the distance increases towards the bulk of substrate material. The central difference scheme is adopted for the diffusion terms. The convergence criterion for the residuals is set as $|\psi^k - \psi^{k-1}| \leq 10^{-6}$ to terminate the simulations.

3.4 Discussions

The discussions on the findings are presented according to the below sub-headings.

3.4.1 Analytical Treatment of Laser Melting Process

The discussions on the solution of analytical approach for the recession velocity are presented in the light of the previous study [2]. Table 3.1 gives the data used in the analysis. Laser non-conduction limited heating is examined and a closed form solution for temperature rise is presented. The recession velocity formulated from energy balance at the surface for steady heating is [3]:

$$V = \frac{I_o}{\rho\left[C_p T_s + L_m\right]} \tag{3.30}$$

where T_s is the surface temperature and L_m is the latent heat of melting. It can be observed that at threshold intensity and the melting temperature, V has a value of about 7 m/s, which is considerably high at the initial stage of the evaporation process. Consequently, recession velocity formulated from energy balance at the surface predicts excessive recession velocity in the early stage of the melting process.

Table 3.1 Material properties used in the simulations. The solidus and liquidus temperatures of steel are 1766 and 1788 K, respectively

Steel			
Temp (K)	*Cp* (J/kgK)	*k* (W/mK)	ρ (Kg/m^3)
300	477	14.9	8,018
400	515	16.6	7968
600	557	19.8	7,868
800	582	22.6	7,769
1,000	611	25.4	7,668
1,200	640	28	7,568
1,500	682	31.7	7,418

3.4.2 Numerical Treatment of Laser Melting

Laser heating of a steel substrate is considered and phase change in the irradiated region is analyzed in line with the previous study [6]. Temperature and flow field in the melt pool are predicted numerically using the control volume approach in line with the experimental conditions. The thermal expansion during the melting process is incorporated through the buoyancy force term in the momentum equation (Eq. 3.19). However, the shrinkage of volume is neglected during the solidification due to rapid cooling. Table 3.1 gives the data used in the simulations.

Figure 3.4 shows temporal variation of temperature at different depths along the symmetry axis and at different locations along the x-axis. Temperature rises sharply in the early heating period and as the heating time progresses temperature rise becomes gradual. Although internal energy gain is high in the surface region, temperature gradient is low due to the attainment of low temperature in the early heating period. This lowers the heat diffusion from the surface region to the solid bulk. Consequently, internal energy gain becomes the dominant mechanism of energy exchange during the early heating period. In the case of the late heating period, the temperature gradient in the surface vicinity of the substrate material becomes high, which, in turn, enhances the heat conduction from the surface. Consequently, the conduction heat transfer from the surface vicinity to solid bulk suppresses the rate of temperature rise in the surface region. Moreover, the rate of temperature rise reduces with increasing depth below the surface. This is particularly true in the early heating period. This is associated with the attainment of low temperature with increasing depth. As the heating progresses ($t > 0.4$ ms), the rate of temperature rise becomes almost similar to all depths below the surface. Figure 3.4 shows temporal variation of surface temperature at different locations along the r-axis while $x = 0$.

Figure 3.5 shows temperature distribution along the r-axis for different heating periods. Temperature decays sharply as the distance along the r-axis increases, particularly onset of the heating period ending (1 ms); in which case, the liquid layer is formed in the surface region. As the distance along the r-axis increases towards the irradiated spot edge, temperature decay becomes gradual. Although the phase change takes place across the melt layer and the solid substrate, steady temperature representing the phase change is not visible. This is attributed to the high heat flux, resembling the laser power intensity, and the close values of liquids and solidus temperatures of the substrate material. It is evident that in the early heating period ($t = 0.25$ ms), temperature gradients attains the local minimum at some locations away from the irradiated spot center. The decay of the temperature gradient towards the local minimum is smooth. However, once the melting initiates and melts layer is formed in the surface region, the temperature gradient curve develops a local peak due to the phase change across the mushy zone, where the liquid and solid phases are co-existed simultaneously. The close values of the solidus and liquidus temperatures as well as the latent heat of melting are responsible for the development of the local peak in the temperature gradient

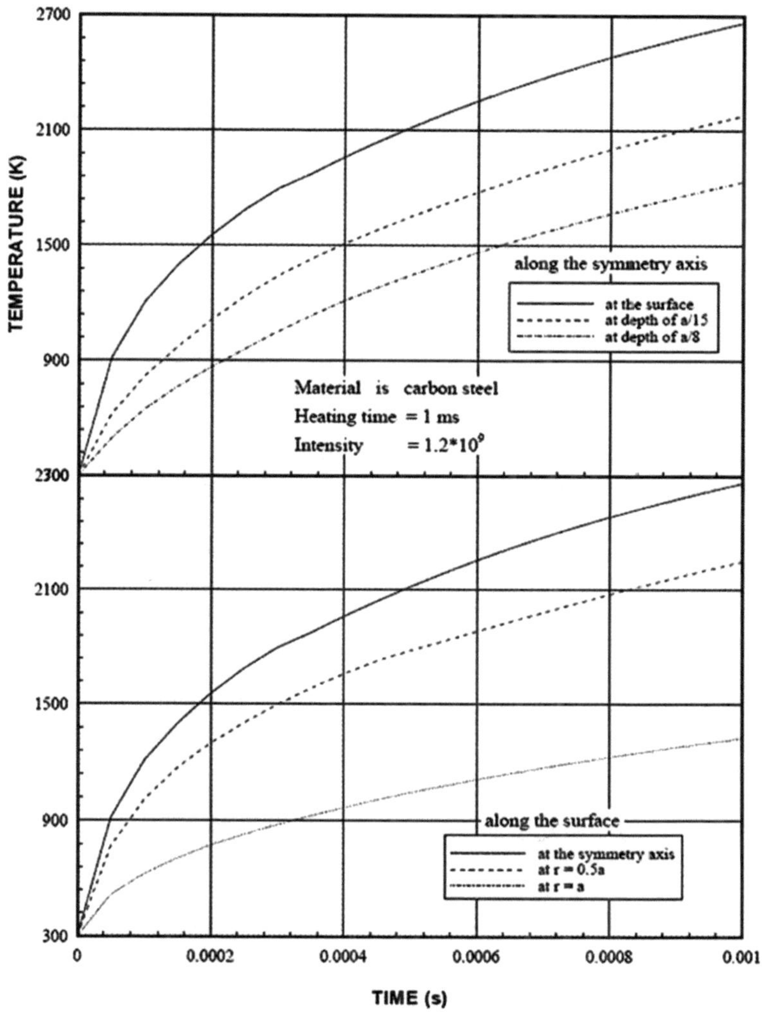

Fig. 3.4 Temporal variation of temperature: **a** along the symmetry axis and **b** along the surface

across the mushy zone. Moreover, the location of the local peak in the temperature gradient moves along the *r*-axis with progressing heating. This is because of the enlargement of the melt pool at the surface of the heated material; in which case, the location of the mushy zone moves away from the irradiated spot center. However, the location of the minimum temperature gradient also moves away from the irradiated spot center with progressing heating duration. This is attributed to the energy gain by the substrate material from the heating source and energy transfer from the surface region to the solid bulk through the axial and the radial conduction. In this case, the internal energy gain from the heat source and the

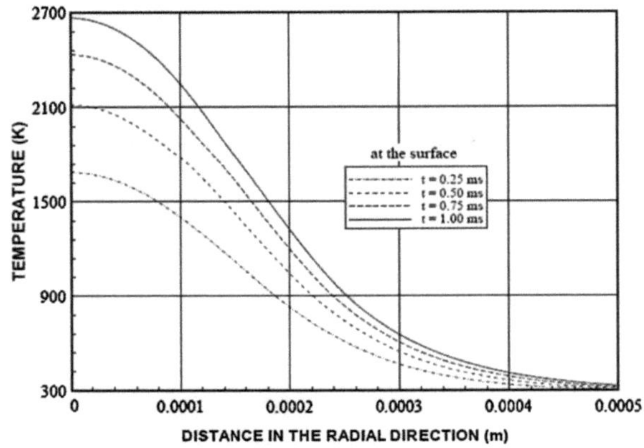

Fig. 3.5 Temperature variation along the radial direction (r-axis) at the surface ($x = 0$) and different heating durations

conduction losses from heated region to the solid bulk are in balance. However, the Gaussian distribution of the laser intensity at the surface modifies temperature field and the temperature gradient variation at the surface. The internal energy gain by the substrate material at the heated surface is not uniform and the conduction losses from the surface region to solid bulk vary due to the spatial variation of the temperature gradient. Consequently, the relation between the location of the minimum temperature gradient and the laser power intensity distribution cannot be simplified.

Figure 3.6 shows temperature distribution along the radial direction for the cases with and without Marangoni effect. It is evident that the Marangoni effect influences temperature distribution in the surface region of the melt pool due to the mixing of the fluid through the convective current generated in this region. Temperature decays sharply in the vicinity of the surface and as the distance towards the solid bulk increases, temperature decay becomes gradual. The rapid decay of temperature is associated with the internal energy gain from the heat surface and the conduction loss due to the temperature gradient. In this case, high

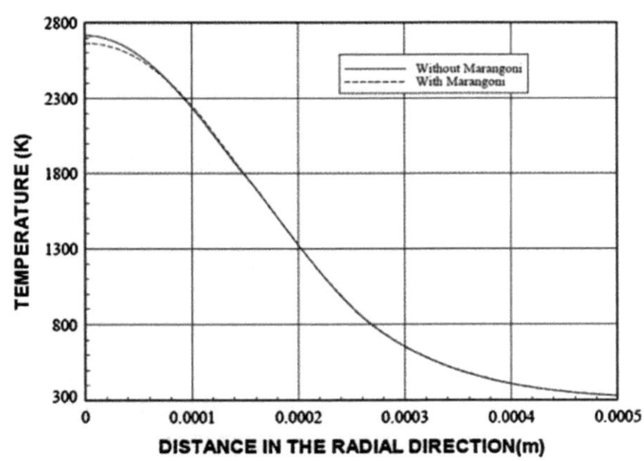

Fig. 3.6 Temperature variation along the radial direction (r-axis) at the surface ($x = 0$) and the heating duration is $t = 1$ ms. Temperature distributions are obtained after considering with and without Marangoni effect

magnitude of the temperature gradient enhances the heat conduction from the surface to its neighborhood. The temperature gradient is in the order of 10^7 K/m, which is significantly higher than that corresponding to its counterpart along the r-axis. Consequently, the local peak formed in the temperature gradient is not visible across the mushy zone. This is because of the small depth of the mushy zone along the x-axis. In reference to the melting temperature of the substrate material, the depth of the melt zone is in the order of 25 μm. Since temperature decay along the x-axis is sharper than that of the r-axis, the enlargement of the melt zone along the x-axis with progressing heating time is much less than that of the r-axis. Although the depth of the melt layer can be increased with increasing the laser pulse intensity, a care must be taken to avoid the evaporation at the surface. Although the evaporation of the liquid metals at the surface depends on the saturation temperature and pressure at the surface, the amount of evaporated liquid metal at liquidus temperature is negligibly small. However, at high laser power intensities, the amount of liquid material evaporated into its ambient become significantly high, which forms the cavity at the surface, i.e., the formation of the cavity becomes unavoidable at high laser power intensities. This, in turn, increases the surface roughness and lowers the applicability of the laser treated surface in practice.

Figure 3.7 shows the maximum velocity in the melt pool with the time. It is evident that in the early melting ($t \leq 0.5$ ms), velocity magnitude is slowly rising. This is mainly because of the shallow melt pool and shear resistance due to the rate of fluid strain, which becomes comparable to the surface tension force. However, as the time progresses, the size of the melt pool increases and the rate of fluid strain, particularly in the melt pool surface region, become small. This, in turn, enhances the velocity magnitude in the circulation cell. The maximum velocity magnitude increases almost linearly with progressing heating time and reaching about 0.15 m/s at the end of 1 ms heating duration.

Figure 3.8 shows the micrograph of scanning electron microscopy (SEM) for the cross-section of the laser melted region of the substrate material. It should be

Fig. 3.7 Temporal variation of the maximum velocity magnitude in the melt pool

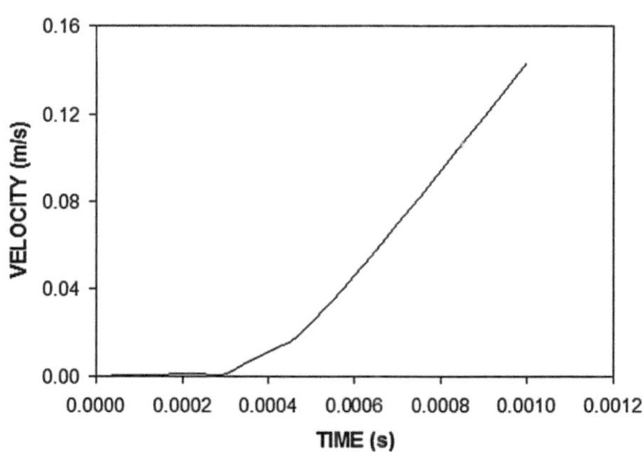

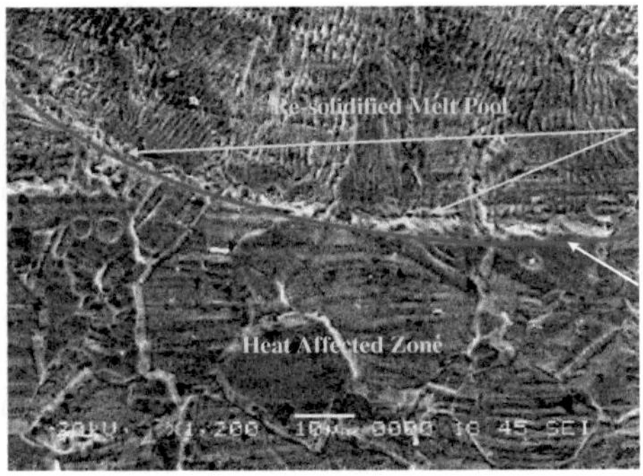

Fig. 3.8 Micrograph of scanning electron microscopy for laser melted region and the melt pool profile

noted that amplified view of the melt pool is shown in the figure in order to observe the melt layer clearly. Consequently, the entire melt pool is not shown in the SEM micrograph. The demarcation zone is visible separating the solid and re-solidified layer in the melt pool. The grains below the re-solidified region remains large indicating the heat affected zone. However, the fine structures are observed in the re-solidified melt pool. This indicates the fast cooling rate in the melt pool. It should be noted that the formation of fine grains results in volume shrinkage in the resolidified region. The melt profile is predicted and also plotted in the SEM micrograph. It is evident that the melt pool profile predicted from the simulations agrees with the experimental results. However, the discrepancies between the both particles are associated with the assumptions made in the simulation, such as uniform properties, shrinkage of volume during rapid resolidification of the melt pool, and the experimental errors, which is estimated as 7 % based on the repeatability of the experiments.

3.4.3 Influence of Pulse Profile and Marangoni Flow on Temperature Field

The discussions on the influence of the laser intensity parameter (β) on the melt pool and the Marangoni flow are presented in line with the previous study [12]. Table 3.2 is used in the simulations.

Figure 3.9 shows temporal variation of temperature at the irradiated spot center for different laser intensity parameters with and without the inclusion of the Marangoni effect in the analysis. It is evident that the influence of the Marangoni effect on the temporal behavior of the surface temperature is notable for the intensity parameter $\beta = 0$. This effect is minimal for the other intensity parameters considered in the simulations. The intensity parameter $\beta = 0$ corresponds to the Gaussian distribution of the laser intensity at the workpiece surface. Although the intensity parameter modifies the spatial distribution of laser power intensity at the workpiece surface, the energy content of each heating situation with different

Table 3.2 Laser pulse parameters used in the simulations

Laser pulse length t_p (μs)	Cooling period t_c (μs)	Pulse rise time t_r (μs)	Pulse fall time t_f (μs)	Power (W)	Pulse parameter β	Gaussian parameter a $(m) \times 10^{-4}$
100	50	1	1	131.947	0–1	2

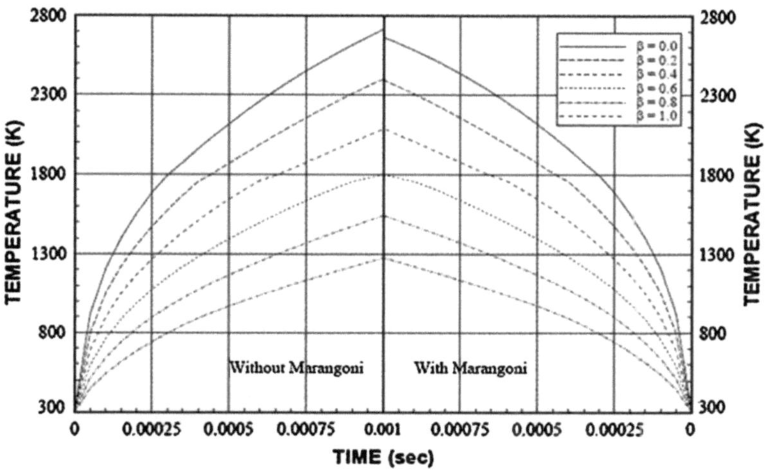

Fig. 3.9 Temporal variation of temperature distribution for different values of intensity parameter (β). The influence of Marangoni effect on temperature distributions is shown for comparison reason

intensity parameters remains the same. The peak laser intensity moves away from the symmetry axis with increasing intensity parameter (β) (Fig. 3.10). Consequently, the high power intensity at the irradiated spot center results in the attainment of the maximum surface temperature in this region. This, in turn, enhances the melt pool depth and lowers temperature at the irradiated spot center due to the Marangoni flow developed in the melt pool, i.e., convection current carries relatively low temperature fluid from the irradiated spot edges and after mixing with the high temperature fluid in the irradiated spot center, temperature reduces in this region. Moreover, the rate of rise of temperature in the early heating period $(t \leq 0.0001\ \text{s})$ is faster than that of the late heating period $(0.0005 \leq t \leq 0.001\ \text{s})$, i.e., the slope of temperature curve $\left(\frac{\partial T}{\partial t}\right)$ is large in the early heating period and it becomes relatively lower in the late heating period as compared to that of the early heating period. This is true for all the values of the intensity parameters considered in the simulations. This is attributed to the internal energy increase in the early heating period and low temperature gradient in the surface region. In this case, energy gain from the irradiated field is high in the surface vicinity and the temperature gradient becomes low in the early heating period. This, in turn, enhances the rate of temperature rise in this region. As the

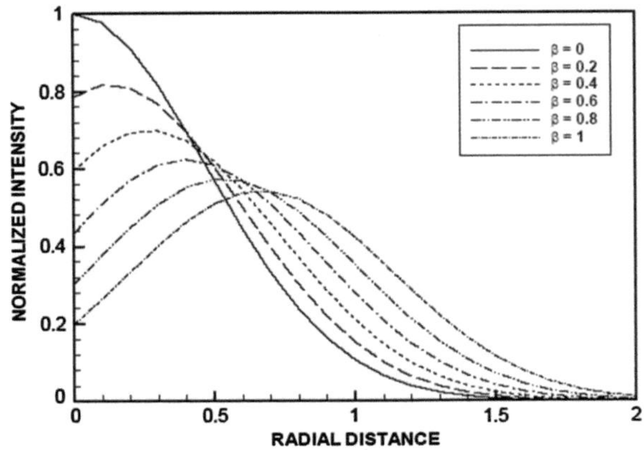

Fig. 3.10 Pulse intensity distribution for different values of intensity parameter (β)

heating period progresses, temperature gradient becomes large in the region next to the surface vicinity. This enhances the energy transfer by heat conduction from the surface region towards the solid bulk. Consequently, the rate of internal energy gain of the substrate in the surface region becomes low. This appears as a low rate of rise of temperature at the surface. As the heating period progresses further, temperature rise becomes gradual and it increases almost at a steady rate.

Figure 3.11 shows temperature distribution along the radial direction at the surface of the workpiece with and without Marangoni effect for different values of the intensity parameters. The influence of the Marangoni effect is notable for the intensity parameter $\beta = 0$. This is again because of the laser peak power intensity, which is located at the center of the irradiated spot due to the Gaussian intensity distribution. Consequently, high surface temperature results in deep melt pool in the surface region and flow mixing due to convective current in the melt pool lowers surface temperature at the irradiated spot center. As intensity parameter reduces, the peak laser power intensity moves away from the irradiated spot center (Fig. 3.10). This, in turn, results in the peak temperature away from the center of the irradiated spot in the radial direction. Since the peak power intensity reduces with increasing power intensity parameter, the peak temperature also reduces with increasing intensity parameter. Consequently, temperature profiles in the radial direction behave similar to the laser intensity profiles along the radial direction. Temperature decay becomes gradual in the region of the peak temperature and as the radial distance increases from this location, temperature decay becomes sharp. The sharp decay of temperature is attributed to the temperature gradient in this region. It should be noted that the high temperature gradients accelerates the heat conduction in the radial direction. Temperature decay becomes gradual as the values of the intensity parameters increases. The size of the constant temperature region in the mushy zone around the melt pool is small, despite the fact that superheating of liquid phase occurs in the melt pool, particularly for $\beta \leq 0.4$. It should be noted that superheating in the melt pool enhances the heat transfer in the radial direction while enhancing the melt pool size along the radial axis.

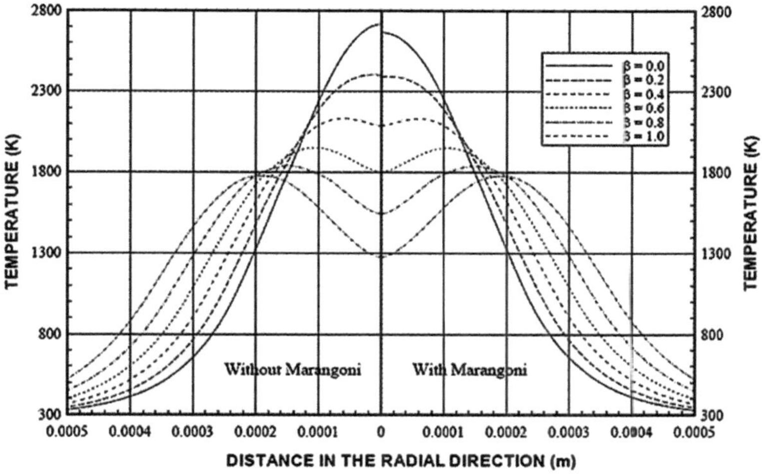

Fig. 3.11 Temperature distribution along the radial direction at the workpiece surface for different values of intensity parameter (β). The influence of Marangoni effect on temperature distributions is shown for comparison reason

Figure 3.12 shows temperature distribution along the symmetry axis inside the substrate material with and without the Marangoni effect inclusion in the analysis for different values of the intensity parameter. The influence of the Marangoni effect is notable for intensity parameter $\beta = 0$; in which case, temperature at the surface becomes less when Marangoni flow is incorporated in the analysis. However, the slope of temperature decay is sharp in the surface vicinity. This may be attributed to the high rate of conduction heat transfer from the surface region to the solid bulk. The superheating in the liquid phase takes place in the melt pool, which, in turn, results in sharp decay of temperature in the melt zone. The size of the mushy zone is narrow so that steady temperature across the mushy zone is not observed. Temperature becomes gradual as the distance increases below the depth $\left(x = 5 \times 10^{-5}\ \text{m}\right)$. This is because of the amount of heat transfer from the irradiated surface to this region, which is not as high as that takes place in the surface region. The low temperature gradient reduces the amount of conduction heating in this region.

Figure 3.13 shows the velocity vectors in the melt pool predicted from the simulations and SEM micrograph of the re-solidified melt pool. It should be noted that the velocity vectors are predicted for the intensity parameter $\beta = 0$, which is the Gaussian intensity mode. Moreover, the close view of the re-solidified melt pool of is shown in the SEM micrograph to observe the melt pool profile. Therefore, the full melt pool could not be shown in the scale of the SEM micrograph. Nevertheless, it is evident that the melt pool profile predicted is almost identical to its counterpart obtained from the experiment. In addition, the demarcation line separating the re-solidified melt pool profile is clearly notable.

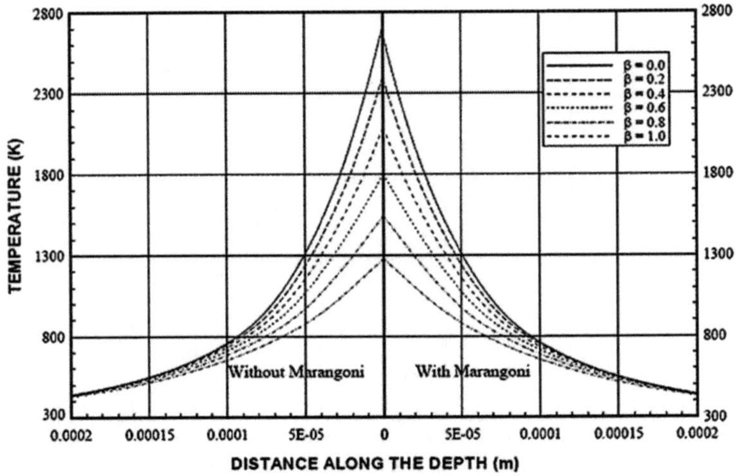

Fig. 3.12 Temperature distribution inside the workpiece for different values of intensity parameter (β). The influence of Marangoni effect on temperature distributions is shown for comparison reason

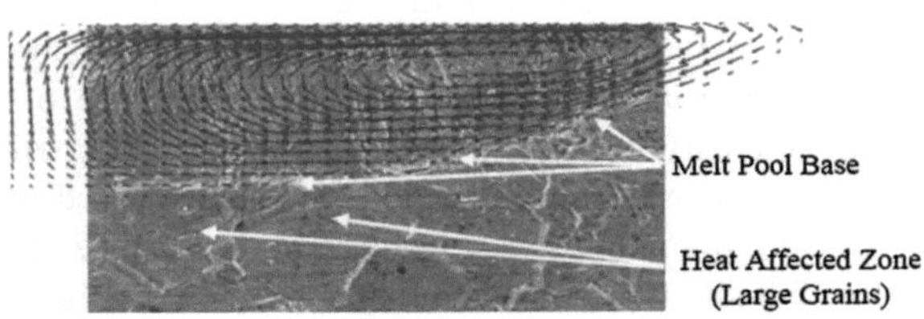

Fig. 3.13 Micrograph of scanning electron microscopy (SEM) for laser melted region and the velocity vectors developed in the melt pool for intensity parameter $\beta = 0$ (Gaussian distribution). SEM micrograph is shown some region of the melt pool due to magnification

The heat affected zone is also notable due to large gains, provided that the size of the heat affected zone is considerably narrow. This is attributed to the high temperature gradients and fast cooling rates in the region next to the melt pool in the solid substrate.

3.4.4 Moving Heat Source Consideration

Laser repetitive pulse heating of titanium alloy is carried out and temperature field and Maragoni flow in the laser irradiated region are modeled numerically. The discussions on the influence of laser scanning speed and laser pulse parameter on temperature field and melt pool size are presented in line with the previous study

Table 3.3 Material properties used in the simulations. The solidus and liquidus temperature of titanium is 1958 K

Titanium								
	Temp (K)	300	400	500	600	800	1,200	2,000
	k (W/mK)	21.9	20.4	19.7	19.4	19.7	22.2	28.7
	Temp (K)	293	373	473	673	873	1,073	1,958
	Cp (J/kgK)	519	540	569	619	636	682	700
	ρ (Kg/m^3)	4,500	4,490	4,478	4,452	4,425	4,397	4,130

[13]. Tables 3.2 and 3.3 give the laser pulse and material properties used in the simulations.

Figure 3.14 shows temporal variation of surface temperature at the irradiated spot center for various laser scanning speeds and two laser pulse parameters. The solid phase heating, melting, and the liquid phase heating are evident for the laser pulse parameter $\beta = 0$. This is true for all laser scanning speeds and laser pulse parameters considered in the simulations. However, workpiece surface does not undergo melting at the irradiated spot centre on the surface for all laser scanning speeds, since temperature remains below the melting temperature of the substrate material. This is associated with the laser pulse intensity distribution at the irradiated spot center, which is less than that corresponding to the Gaussian distribution ($\beta = 0$). Moreover temperature difference between the maximum and the minimum temperatures is considerably small for $\beta = 1$ as compared to that of $\beta = 0$ due to laser power intensity at the irradiated spot center. However, the phase change is evident for $\beta = 0$; in which case temperature remains almost the same with time during the phase change. The liquid phase heating results in further increase in temperature from the melting temperature in the pulse. This appears as a temperature peak at the surface during the heating pulse. However, melting and liquid phase heating starts within the fourth pulse. As the scanning speed increases, the temporal location of liquid phase heating moves further towards the pulse ending. As the number of pulses increase temperature at the surface becomes higher than the melting temperature during the heating and cooling cycle of each pulse. This takes place after the 7th pulse for laser scanning speed 0.1 m/s. However, this situation occurs at different pulse number for laser scanning speed less than $u \leq 0.25$ m/s. Melting, liquid phase heating, and subsequent solidification takes place after the 7th pulse for laser scanning speed 0.30 m/s. The close examination of rate of temperature rise and fall for the last pulse (20th pulse) reveals that the ratio of temperature rise is slower than the rate of fall. This is attributed to the liquid phase heating and heat transfer from the liquid zone to the solid bulk in the laser irradiated region, which is higher in the heating pulse as compared to the cooling period following the pulse.

Figure 3.15 shows temperature distribution along the *x*-axis, for two laser scanning speeds and two laser pulse parameters at the end of the 20th heating pulse. It should be noted that *x* and *y*-axis locations are at the center of irradiated spot. Temperature attains well above the melting temperature of the substrate

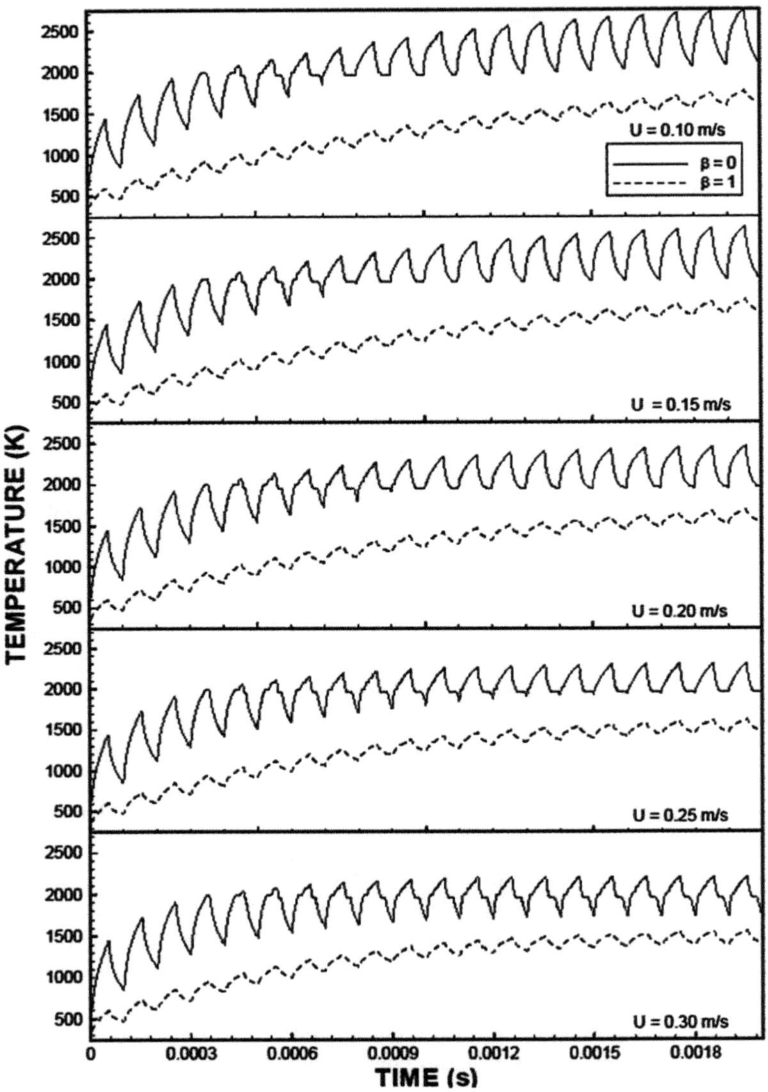

Fig. 3.14 Temporal variation of temperature at the irradiated spot *center*

material at the surface for $\beta = 0$; in which case, the laser pulse intensity distribution at the surface is Gaussian. The liquid phase heating takes place in the vicinity of the surface and the melt region extends almost 50 μm below the surface. Temperature gradient changes within the vicinity of the melt pool wall where temperature reduces to the melting temperature of the substrate material. In this case, temperature decay is sharper in the melt pool as compared to the region

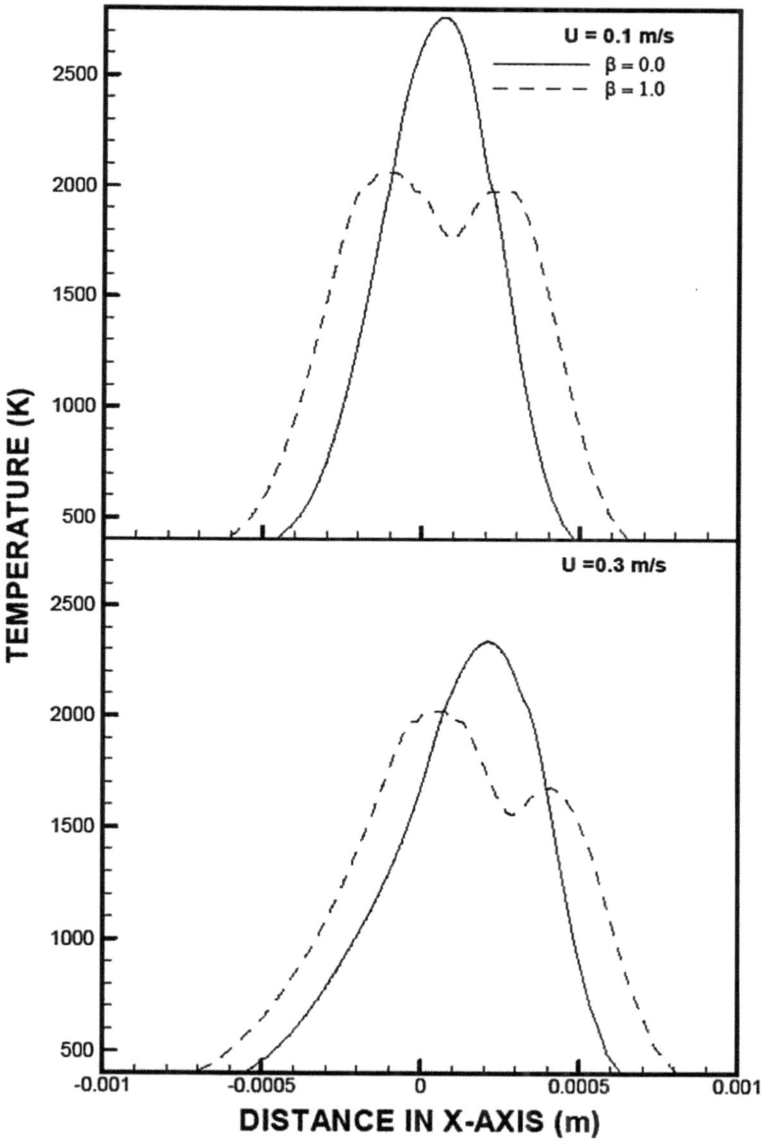

Fig. 3.15 Temperature variation along the x-axis at the end of the 20th pulse

corresponding to next to the melt pool. The sharp decay of temperature in the melt pool is associated with absorption of the laser energy from the irradiated field, which decays exponentially with increasing depth according to the Lambert and Beer's law. However, increasing laser scanning speed reduces the peak temperature at the surface. In addition, the melt depth also reduces considerably in the

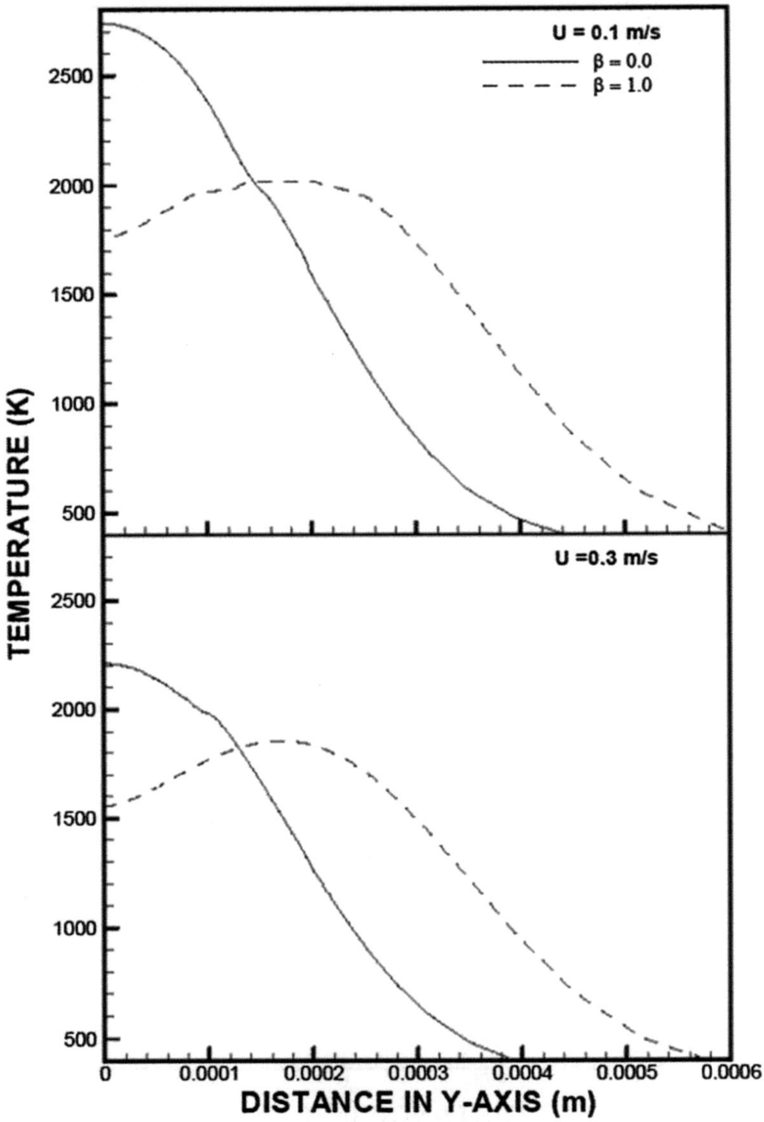

Fig. 3.16 Temperature variation along the y-axis at the end of the 20th pulse

surface region of the workpiece. Temperature decay in the solid phase decreases slightly with increasing depth below the surface. It should be noted that heat conduction increases with increasing temperature gradient and heat conduction governs the energy transfer in the solid phase of the substrate material below the melt pool. Therefore, heat conduction towards the solid bulk reduces with increasing depth at the end of 20th heating pulse. In the case of the laser pulse parameter $\beta = 1$, the peak temperature at the surface reduces significantly due to low laser power intensity at the irradiated spot center. The peak temperature reduces further with increasing laser scanning speed (0.3 m/s) because of the rate of energy deposited into the substrate material which reduces with increasing scanning speed.

Figure 3.16 shows temperature distribution along the y axis for two laser scanning speeds and two laser pulse parameters. The x and z axes locations correspond to the irradiated spot center. The melt size along the y axis is notable for the laser pulse parameter $\beta = 0$ for two laser speeds, i.e., temperature remains almost the same across the solid and the melt zone where the mushy zone is generated. The width of the melt pool along the y axis is the order of 180 μm for low laser scanning speed (0.1 m/s), but it is 100 μm for the laser scanning speed of 0.3 m/s for $\beta = 0$. It should be noted that the laser beam diameter at the surface is about 600 μm. Due to the Gaussian distribution of the laser beam intensity at the surface, almost 2/3rd of the laser irradiated spot is melted for the laser scanning speed 0.1 m/s while 1/3rd of the laser irradiated spot is melted for scanning speed 0.3 m/s. This indicates that the influence of the laser scanning speed has significant effect on the melt width at the workpiece surface. Temperature decay in the irradiated spot center follows almost the Gaussian distribution because of the laser pulse intensity distribution at the surface for $\beta = 0$. In the case of $\beta = 1$, temperature profile follow almost the laser pulse intensity distribution at the surface (Fig. 3.10). However, at low laser scanning speed (0.1 m/s), melting occurs at the surface where laser intensity is high. However, the liquid phase heating in the melt pool is unlikely, since the peak temperature at the surface is slightly higher than the melting temperature of the substrate material.

References

1. Low, D.K.Y., Li, L., Byrd, P.J.: Hydrodynamic physical modeling of laser drilling, ASME. J. Manuf Sci Eng **124**, 852–862 (2002)
2. Yilbas, B.S., Shuja, S.Z.: Laser non-conduction limited heating and prediction of surface recession velocity in relation to drilling. In: Proceedings of Institution of Mechanical Engineers Part C: J. Mech. Eng. Sci. **217**, 1067–1076 (2003)
3. Yilbas, B.S., Sahin, A., Davies, R.: Laser heating mechanism including evaporation process initiating the laser drilling. Int. J. Mach. Tools Manuf. **35**, 1047–1062 (1995)
4. Yilbas, B.S., Sahin, A.Z.: Turbulent boundary layer approach allowing chemical reactions for CO2 laser oxygen-assisted cutting process. In: Proceedings of Institution of Mechanical Engineers Part C: J. Mech. Eng. Sci. **208**, 275–284 (1994)
5. Yilbas, B.S., Davies, R., Gorur, A., Yilbas, Z., Begh, F., Akcakoyun, N., Kalkat, M.:Investigation into development of liquid layer and formation of surface plasma during CO_2 laser cutting process. In: Proceedings of Institution of Mechanical Engineers, Part B: J. Eng. Manuf. **206**, 287–298 (1992)
6. Shuja, S.Z., Yilbas, B.S.: Laser heating and flow field developed in the melt pool. Numer. Heat Transf. Part A **59**(12), 970–987 (2011)

7. Amada, B.P.: 41040, 95912 Roissy Aeroport Cedex, France
8. Shuja, S.Z., Yilbas, B.S.: Pulsative heating of surfaces. Int. J. Heat Mass Transf. **41**, 3899–3918 (1998)
9. Fluent Inc 2005 Fluent Users Guide
10. Voller, V.R., Prakash, C.: A fixed-grid numerical modeling methodology for convection-diffusion mushy region phase-change problems. Int. J. Heat Mass Transf. **30**, 1709–1720 (1987)
11. Patankar, S.V.: Numerical Heat Transfer. McGraw-Hill, New York (1980)
12. Shuja, S.Z., Yilbas, B.S., Momin, O.: Laser repetitive pulse heating and melt pool formation at the surface. J. Mech. Sci. Technol. **25**(2), 479–487 (2011)
13. Shuja, S.Z., Yilbas, B.S., Momin, O.: Laser heating of a moving slab: Influence of laser intensity parameter and scanning speed on temperature field and melt size. Opt. Lasers Eng. **49**(2), 265–272 (2011)

Chapter 4
Laser Melting of Two Layer Materials

Abstract Laser melting of surfaces results in homogeneous structures in the laser treated layer, which is required for the practical applications of coated surfaces. In addition, laser control melting of ceramic coatings provides thermal integration between the coating and the base substrate through generating strong bonding of the coating to the base material. The differences in the thermal properties of the coating and the base material, and the small coating thickness result in differences in the size of the melting zones in the coating and in the substrate material. Laser re-melting of surfaces is involved with the Marangoni flow in the melt pool. This is because of the surface tension force gradient acting at the free surface of the melt pool. Investigation of the laser melting of two-layer assembly, consisting of thin layer (≈ 10 μm) at the top and the base material at the bottom, including the effects of Marangoni flow, coating thickness, and different coating materials provides physical insight into the melting of coated material. In this chapter, laser melting of two layer assembly is presented in line with different heating and boundary conditions including the stationary and different laser pulse intensity distributions.

4.1 Introduction

Laser melting of surfaces results in homogeneous structures in the laser treated layer, which is required for the practical applications of coated surfaces. In addition, laser control melting of ceramic coatings provides thermal integration between the coating and the base substrate through generating strong bonding of the coating to the base material. The differences in the thermal properties of the coating and the base material, and the small coating thickness result in differences in the size of the melting zones in the coating and in the substrate material. Laser re-melting of surfaces is involved with the Marangoni flow in the melt pool. This is because of the surface tension force gradient acting at the free surface of the melt pool. Investigation of the laser melting of two-layer assembly, consisting of thin

B. S. Yilbas and S. Z. Shuja, *Laser Surface Processing and Model Studies*, Materials Forming, Machining and Tribology, DOI: 10.1007/978-3-642-36629-1_4,

layer ($\approx$10 μm) at the top and the base material at the bottom, including the effects of Marangoni flow, coating thickness, and different coating materials provides physical insight into the melting of coated material.

4.2 Numerical Treatment of Melting Process

Since the multi-physics, including heat transfer and fluid flow, are involved during laser melting of two-layer materials, analytical solution to the problem becomes very difficult to achieve with accuracy. Therefore, the numerical solution of the problem remains unavoidable alternative. The analysis related to the phase change during laser heating is provided in Chap. 3, only the brief formulation and boundary conditions are given herein. The model study is presented in the light of the previous studies [1–3].

Figure 4.1 shows the schematic view of laser melting of two-layer structure. Since the heated substrate material is stationary, the convective terms drop and governing equation reduces to:

$$\rho \frac{\partial (CpT)}{\partial t} = (\nabla (k \nabla T)) + S_o \tag{4.1}$$

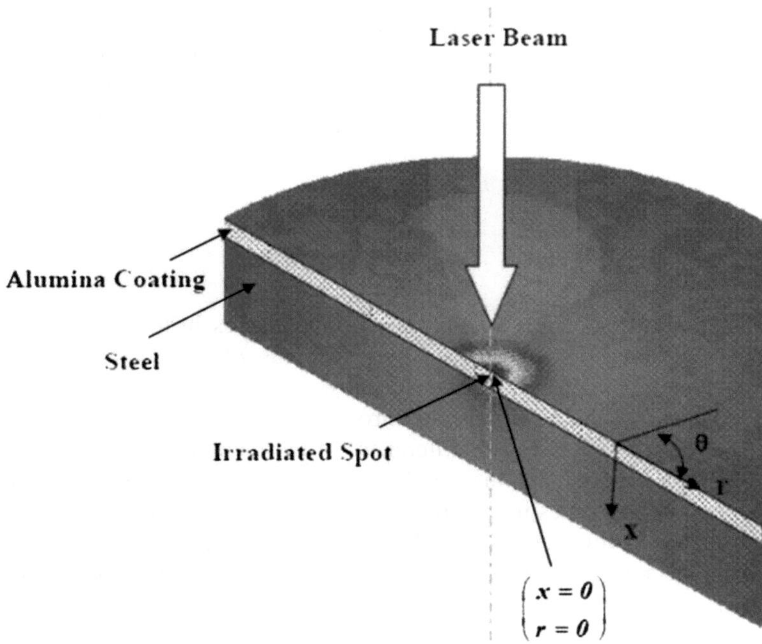

Fig. 4.1 A schematic view of the laser heating of alumina coated steel

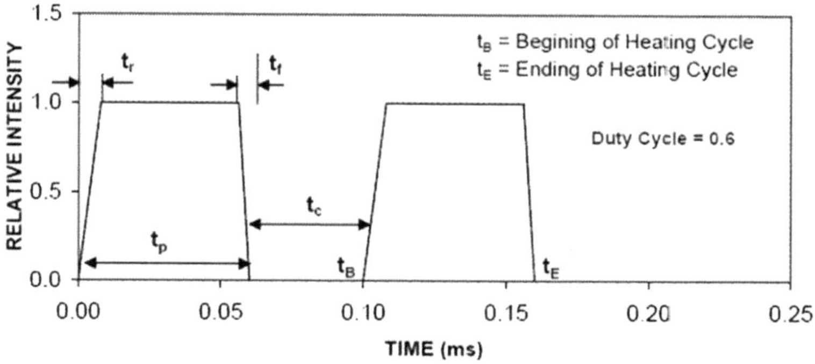

Fig. 4.2 Temporal variations of laser repetitive two pulses

where ρ is the density, Cp is the specific heat, k is the thermal conductivity, and S_o is the volumetric source resembling the laser power absorption in the substrate material, i.e.,:

$$S_o = I_o\delta(1 - r_f)\exp(-\delta x)\exp\left(-\left(\frac{r}{a} + \beta\right)^2\right)f(t) \tag{4.2}$$

I_o is laser peak intensity, δ is the absorption depth, r_f is the surface reflectivity, $f(t)$ is the temporal distribution of the laser repetitive pulses (Fig. 4.2), a is the Gaussian parameter, and β is the laser intensity parameter. The temporal variation of laser pulse intensity is considered to be in trapezium shape in time domain. The temporal variation of the laser pulse shape, which is trapezium in time domain, resembles almost the actual laser pulse shape used in the industry. The time function ($f(t)$, Fig. 4.2) representing the consecutive pulses is:

$$f(t) = \left\{\begin{matrix} 0, & t = 0 \\ 1, & t_r \leq t \leq t_f \\ 0, & t = t_p \\ 0, & t_p \leq t \leq t_c \end{matrix}\right\} \tag{4.3}$$

where t_r is the pulse rise time, t_f is the pulse fall time, t_p is the pulse length, t_c is the end of cooling period. $f(t)$ repeats when the second consecutive pulse begins, provided that time $t = t_f + t_c$ corresponds to the starting time of the second pulse. The same mathematical arguments can apply for the other consecutive pulses after the second pulse.

In the case of solid heating, two boundary conditions for each principal axis are specified. Due to the low absorption depth, at a distance considerably away from the surface (at infinity) it is assumed that the heating has no effect on the temperature of the slab; consequently, at a depth of infinity, the temperature is assumed to be constant and equal to the initial temperature of the slab. However, at alumina-steel interface the continuity of flux and temperature is incorporated. The boundary conditions, therefore, are:

x at infinity	$x = \infty$	$\Rightarrow$	$T(r, \infty, t) = T_0(\text{specified})$
r at infinity	$r = \infty$	$\Rightarrow$	$T(\infty, x, t) = T_0(\text{specified})$
At symmetry axis	$r = 0$	$\Rightarrow$	$\frac{\partial T(0,x,t)}{\partial r} = 0$
At the surface	$x = 0$	$\Rightarrow$	$k\frac{\partial T(r,0,t)}{\partial x} = h_t(T_s - T_\infty) + \sigma_t(T_s - T_\infty)$

where h_t is the heat transfer coefficient at the free surface and σ is the Stefan-Boltzmann constant. The heat transfer coefficient predicted earlier is used in the present simulations $\left(h_t = 10^4\ \text{W/m}^2\text{K}\right)$

At the alumina-steel interface:

$$x = x_t : k_a \frac{\partial T}{\partial x} = k_s \frac{\partial T}{\partial x}$$

where k_a and k_s are the thermal conductivity of alumina and steel, respectively, and x_t is the alumina coating thickness (10 μm).

The laser beam axis is the z-axis (Fig. 4.1). Equation 4.1 is solved numerically with the appropriate boundary conditions to predict the temperature field in the substrate material. However, to analyze the phase change problem, the enthalpy-porosity technique is used. In this case, the melt interface is tracked explicitly after defining a quantity called the liquid fraction, which indicates the fraction of the cell volume that is in liquid form. Based on the enthalpy balance, the liquid fraction is computed. The mushy zone is a region in which the liquid fraction lies between 0 and 1. The mushy zone is modeled as a "pseudo" porous medium in which the porosity decreases from 1 to 0 as the material solidifies. When the material has fully solidified in a cell, the porosity becomes zero and hence the velocities also drop to zero.

To discretize the governing equations, a control volume approach is introduced. The details of the numerical scheme are given in Chap. 3. The calculation domain is divided into grids and a grid independence test is performed for different grid sizes and orientation. A non-uniform grid with 90 × 55 mesh points along x and r-axes, respectively, is employed after securing the grid independence. The finer grids are located near the irradiated spot center in the vicinity of the surface and grids become courser as the distance increases towards the bulk of substrate material. The central difference scheme is adopted for the diffusion terms. The convergence criterion for the residuals is set as $\left|\psi^k - \psi^{k-1}\right| \leq 10^{-6}$ to terminate the simulations. The laser energy is kept the same for all intensity distributions with different intensity parameters.

4.3 Discussions

Laser melting of two-layer structures and numerical modeling is presented in the light of the previous studies [1–3] under the appropriate sub-headings.

4.3.1 *Laser Melting of Two-Layer Structure Influence of Laser Pulse Repetition on Temperature and Flow Field in the Melt Pool*

The findings for the laser repetitive pulse heating of surface of the aluminum coated steel are presented in the light of the previous study [1]. The alumina coating has 10 μm thickness at the steel surface. The phase change in alumina and in the surface region of the steel is incorporated in the analysis. Tables 4.1 and 4.2 give the laser parameters and material properties incorporated in the simulations.

Figure 4.3 shows temporal variation of temperature at the surface of alumina at alumina-steel interface and at some depth inside steel for 10 laser pulses. It should be noted that each laser pulse has heating and cooling periods due to the duty cycle, which is 60 % i.e., 0.6^*t_p is the heating and 0.4^*t_p is the cooling period, t_p being the pulse length. Temperature rises rapidly at the surface and at the other locations below the surface, provided that the rise of temperature is high at the surface. This is attributed to the low thermal conductivity of alumina as compared to steel. In addition, the laser power intensity is the maximum at the surface. The melting temperature of steel is well below the melting temperature of alumina (Table 4.1); therefore, steel undergoes a melting after the second pulse while alumina remains in the solid phase. Since the phase change in the steel surface takes place rapidly, attainment of constant temperature during the heating cycle is not notable from temperature curve. However, the decay rate of temperature

Table 4.1 Laser pulse parameters used in the simulations

Duty cycle (%)	Laser pulse length t_p (ms)	Cooling period t_c (ms)	Pulse rise time t_r (ms)	Pulse fall time t_f (ms)	Power intensity $(W/m)^2$	Guassian parameter a (m)
60	0.06	0.04	0.0078	0.0039	0.6×10^{10}	3×10^{-4}

Table 4.2 Material properties used in the simulations [4]

Steel	*Temp* (K)	300	400	600	800	1,000	1,200	1,500
	Cp (J/kgK)	477	515	557	582	611	640	682
	k (W/mK)	14.9	16.6	19.8	22.6	25.4	28	31.7
	$\rho(kg/m^3)$	8,018	7,968	7,868	7,769	7,668	7,568	7,418
	$T_{solidus} = 1766$ K		$T_{biquidus} = 1788$ K			$L = 400{,}000$ J/kg		
Alumina	*Cp* (J/kgK)	786	940	1020	1,140	1,191	1,234	1,255
	k (W/mK)	37.1	28.2	21.8	11.6	8.6	7.1	6.7
	$\rho(kg/m^3)$	3,800						
	$T_{solidus} = 2260$ K		$T_{biquidus} = 2323$ K			$L = 900{,}000$ J/kg		

The solidus and liquidus temperatures of steel are 1,766 and 1,788 K while solidus and liquidus temperatures of alumina are 2,260 and 2,323 K, respectively. The latent heat of melting for steel is 400,000 J/kg and for alumina is 900,000 J/kg

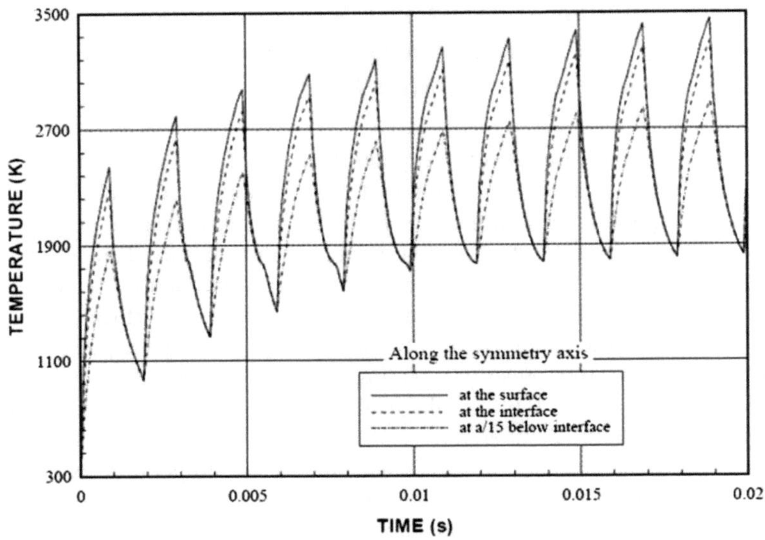

Fig. 4.3 Temporal variation of temperature at different depths

changes in the cooling cycle at the interface and at a depth $a/15$ (6.7×10^{-5} m) below the surface. The change of temperature decay rate in the cooling period of the pulse indicates the presence of the phase change during and after the second pulse. As the time progresses further, temperature increases at the surface and reaches the melting temperature of alumina. In this case, temperature increase changes indicating the initiation of the phase change at the alumina surface. This is more pronounced as the number of the laser pulses increases. Moreover, temperature at interface and at depth $x = 10^{-5}$ m below the interface remains in liquid phase during the 6 and more pulses, since temperature remains above the melting temperature of steel in this region.

Figure 4.4 shows temperature variation along the radial direction at the surface for the beginnings and endings of 9th and 10th pulses. It should be noted that the laser pulse intensity distribution is Gaussian at the irradiated surface. Consequently, the peak temperature occurs at the irradiated spot center. Temperature reaches its maximum at end of the heating period of the 10th pulse and the difference between the peak temperatures due to 9th and 10th pulses is small. However, temperature difference at the end of the cooling periods of 9th and 10th pulses is negligibly small. Consequently, attainment of the similar distributions of temperature during the heating and cooling cycles prevail that almost steady heating situation is attained after the 9th pulse at the irradiated surface. Temperature decay in the vicinity of the irradiated spot center is lower than that corresponding to next to the surface vicinity. The slow decay of temperature in the irradiated surface is associated with the laser power intensity distribution at the surface. The Gaussian intensity distribution results in the similar temperature

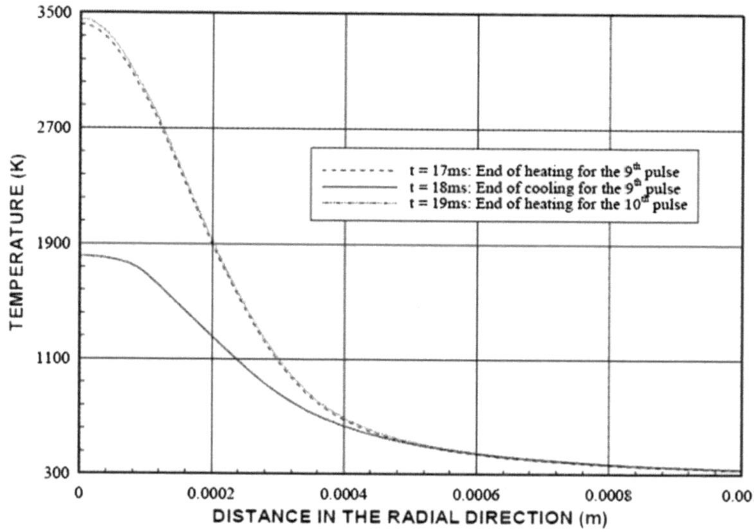

Fig. 4.4 Temperature distribution along the radial direction for different times

distribution at the surface. The phase change, taking place between the solidus and liquidus temperatures, results in almost steady temperature in the radial direction. However, this is not observed at the end of the heating cycle of 9th and 10th pulses. This indicates that the mushy zone, where temperature changes within solidus and liquidus limit, is considerably narrow in the radial direction at the end of the heating period of 9th and 10th pulses. Temperature in the region of the irradiated spot center remains well above the melting temperature of alumina. This shows the super heating in the liquid phase at the alumina surface. However, the region where the superheating takes place is limited to $y = 1.5 \times 10^{-3}$ m from the irradiated spot center. In the case of the end of the cooling period of 9th and 10th pulses, temperature reduces below the melting temperature of alumina. In this case, solid phase heating takes place at the surface. Since the liquid phase occurs in the region of the irradiated spot center, temperature decay along the radial direction in this region is gradual at the end of the cooling period. It should be noted that heat conduction from the superheated liquid to underneath solid surface and heat convection from the surface in radial direction result in almost steady temperature decaying in this region.

Figure 4.5 shows temperature distribution inside the coating and steel for the periods corresponding to the beginning and ending of 9th and 10th pulses. Temperature decays sharply at the end of the heating cycle of 9th and 10th pulses. In addition, difference in temperature profiles at the end of the heating cycle due to both pulses is significantly small. The mush zones in the alumina coating and in the steel are not observable from temperature curves. This indicates that the size of the mushy zone is significantly narrow in the depth of both alumina coating and

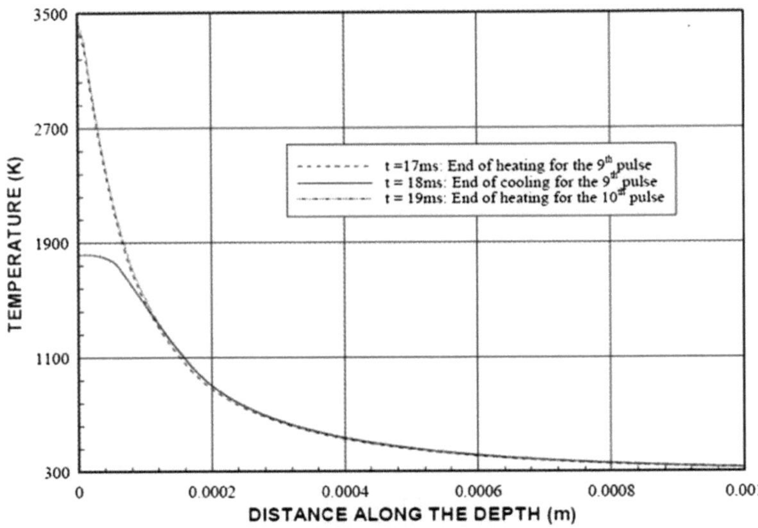

Fig. 4.5 Temperature distribution inside the substrate material for different times

steel. In the case of the end of the cooling cycle of 9th and 10th pulses, both temperature profiles become almost similar along the depth. Temperature in the alumina coating remains below the melting temperature; therefore, it remains in the solid phase. However, temperature in the steel is above the melting temperature at the interface as well as some depth below the steel surface. This results in almost flat temperature distribution along some depth in steel. Consequently, alumina coating remains in the solid phase while steel surface remains in the liquid phase. As the depth below the steel surface increases temperature decays first gradually and later sharply. The gradual decay of temperature is associated with the superheating of liquid phase in steel. Moreover, the superheating in the liquid phase in steel surface acts as a thermal barrier for heat conduction from the alumina coating. This, in turn, results in the attainment of almost steady temperature in the alumina coating.

Figure 4.6 shows temperature contours and liquid phase, mushy zone and solid phase at the end of cooling and heating periods of 9th and 10th pulses. The size of the melt pool (width and depth) is larger during the end of the heating period. This remains almost the same for 9th and 10th pulses. In the case of the end of the cooling period, the width of the melt pool reduces more than that corresponding to the melt depth. In addition, alumina coating remains in the solid phase while the melt pool is present in the region of the steel surface beneath the alumina coating. The size of the mushy zone increases at the end of the cooling period of both pulses. This is because of the heat transfer from liquid phase to solid phase at liquid–solid interface. In this case, the high latent heat of melting of steel is responsible for the increase in the mushy zone size in steel. The melt pool and the

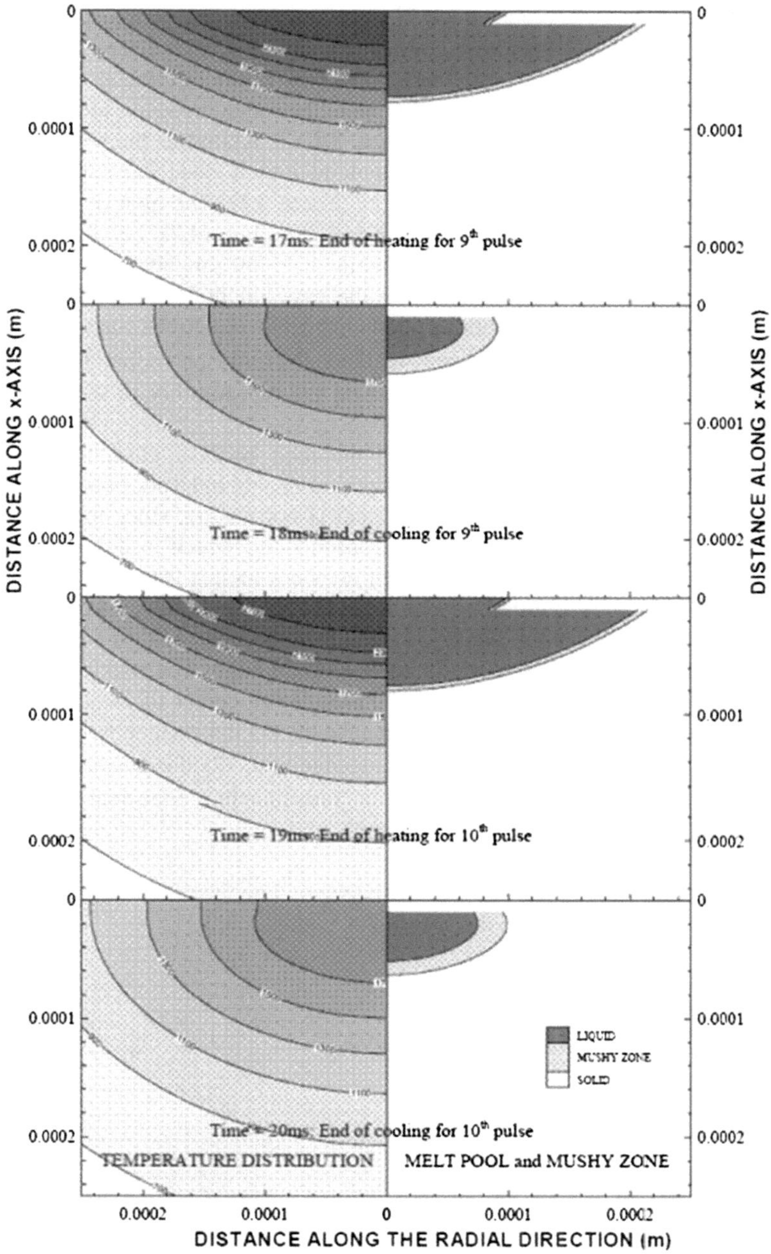

Fig. 4.6 Temperature contours, liquid and solid phases, and mushy zone

mushy zone geometry become round in steel towards the alumina-steel interface. This occurs because of the heat conduction in the radial direction, since the temperature gradient is high in the radial direction in the interfacial region during the cooling period. The size of the melt zone in the radial direction in steel is larger than that corresponding to alumina coating. This is related to the solidus and the liquidus temperatures of alumina and steel; in which case, the solidus and the liquidus temperature are higher for alumina (Table 4.1). The melt depth predicted agrees with the experimented data as depicted from Fig. (4.7), i.e., the melt depth predicted after 10th pulse is 72.5 μm while it is 80 μm corresponding to the experiment. The difference is because of the experimental error, which is in the order of 8 % and the assumptions made in the simulations such as uniform properties and assumption of constant surface reflectivity $\left(r_f = 0.62\right)$. It should be noted that the experimental error is based on the repeatability of the results corresponding three repeats.

Figure 4.8 shows velocity vectors in the melt pool of alumina due to the Marangoni flow at the end of the heating period of 9th and 10th pulses. However, the velocity vectors in the melt pool of steel is not visible, since the flow in the melt pool is governed by the density variation only, which is not significant. It should be noted that the melt pool in steel does not have a free surface; therefore the surface tension is omitted in the analysis. Consequently, Marangoni flow replaces with the flow due to density variation in the melt pool of steel. The flow field forms a circulation cell towards the melt pool edge. The maximum velocity magnitude is in the order of 0.4 m/s. The circulation center of the cell is shifted towards the melt pool surface. This is because of: (1) the surface tension gradient $\left(\frac{\partial \sigma}{\partial T}\right)$, which is high towards the melt pool edge and, (2) the rate of fluid strain, which is high at the melt pool bottom, i.e., resulting in a frictional force, which shifts the cell away from the melt pool wall.

Fig. 4.7 The maximum depth of melt for each pulses predicted and obtained from the experiment

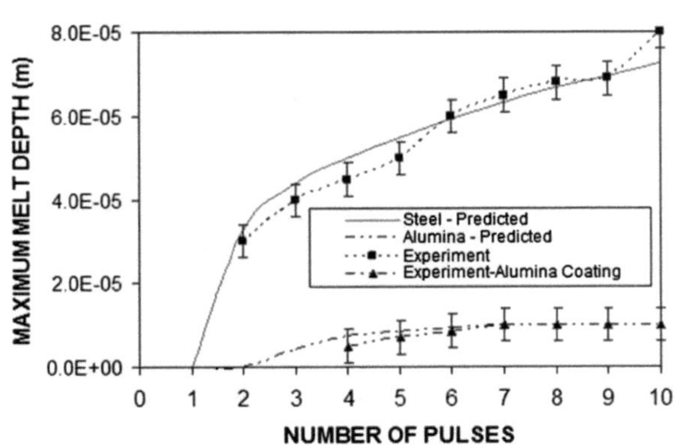

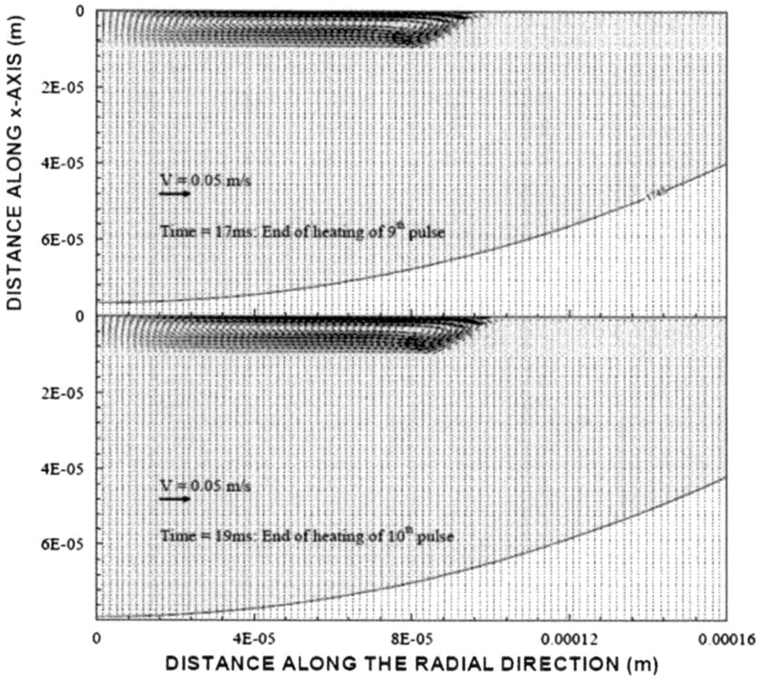

Fig. 4.8 Velocity vectors in the melt pool for different times

4.3.2 Influence of Coating Material on Temperature and Flow Field in the Melt Pool

Carbides and Ceramics layers/coatings on the metallic substrate surfaces can be treated by a laser to generate fine structures in the surface region of the metallic substrates. However, thermal response of the surface to the laser heating pulse changes for different coating materials. Therefore, laser induced melting of coated steel is simulated and the influence of coating material on the melt pool characteristics are presented in the line with the previous study [2]. Alumina, tungsten carbide, and boron are incorporated as coating material in the simulations. The study is extended to include the influence of laser intensity parameter on the temperature field and melt pool size. Tables 4.2, 4.3 and 4.4 give the properties and the laser parameters used in the simulations.

Figure 4.9 shows temperature distribution along the depth (x-axis) for two different time periods, three coating materials and laser intensity parameter $\beta = 0$. Temperature attains significantly high value at the surface (x = 0 m) at the end of the heating cycle of the 10th consecutive pulse (19 ms) for boron, then follows alumina and tungsten carbide. The attainment of high surface temperature is

Table 4.3 Material properties used in the simulations [4]

Tungsten	$C_P = 327$ J/kgK		$K = 70.6$ W/mK			$\rho = 13101$ kg/m^3	
Carbide	$T_{solidus} = 2925$ K		$T_{biquidus} = 2925$ K			$L = 192{,}000$ J/hg	
Boron	$Temp$(K)	300	400	600	800	1,000	1,200	1,500
	Cp(J/kgK)	600	1,463	1,892	2,160	2,383	2,400	2,400
	k(W/mK)	55.5	16.8	10.6	9.6	9.85	9.85	9.85
	ρ(kg/m^3)	2,500						
	$T_{solidus} = 2573$ K		$T_{biquidus} = 2600$ K			$L = 52{,}000$ J/kg		

Table 4.4 Laser pulse parameters used in the simulations for $\beta = 0$

Duty cycle (%)	Laser pulse length t_p (ms)	Cooling period t_c (ms)	Pulse rise time t_r (ms)	Pulse fall time t_f (ms)	Power intensity (W/m)^2	Guassian parameter a (m)
50	1	1	0.156	0.07	0.6×10^{10}	3×10^{-4}

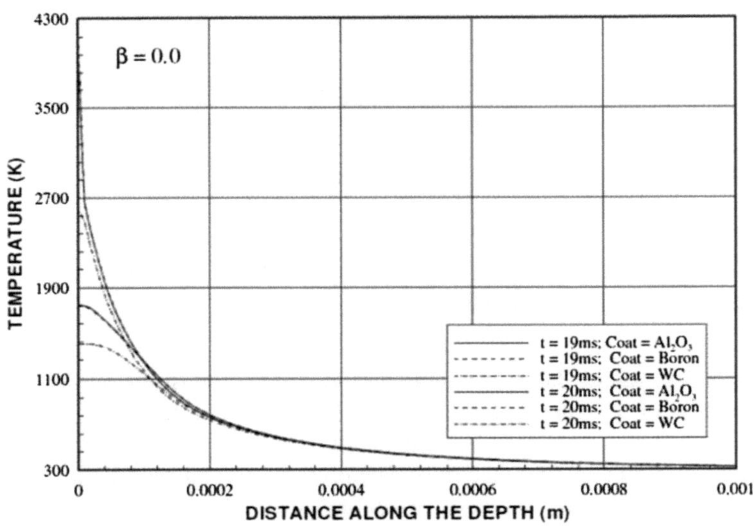

Fig. 4.9 Temperature distribution inside the substrate material along the depth for laser pulse intensity parameter $\beta = 0$, two heating periods and different coating materials

associated with the thermal diffusivity and absorption coefficient of boron and alumina which is lower than tungsten carbide (Table 4.3). In addition, temperature decay is sharp in the surface vicinity, which in turn results in high temperature gradients in this region. This is particularly true for the end of the heating cycle of the 10th pulse. Since temperature exceeds the melting temperature of the coating material along the depth of the coating, which is 7.5 μm. Since temperature below

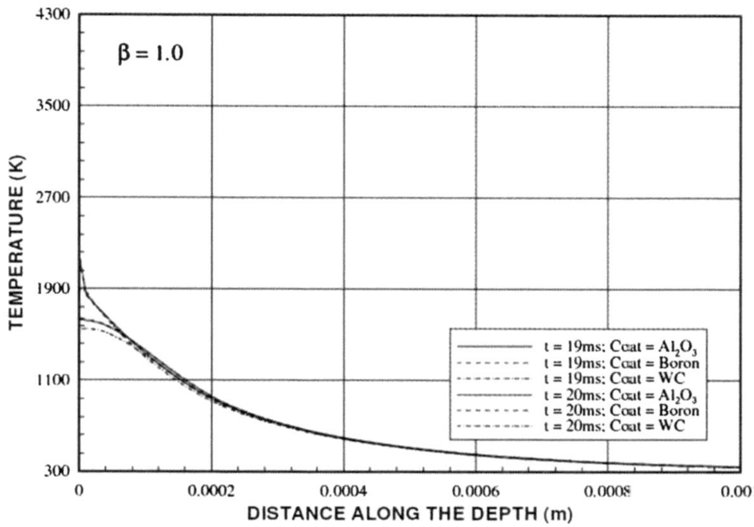

Fig. 4.10 Temperature distribution inside the substrate material along the depth for laser pulse intensity parameter $\beta = 0$, two heating periods and different coating materials

the coating is higher than the melting temperature of steel, melting also takes place in steel in the region close to the interface. Moreover, temperature decay becomes gradual beyond the depth $x \geq 75$ μm and temperature profiles in steel due to different coating materials become almost the same. It should be noted that the size of the mushy zone between the liquid and solid phases is not visible from the figure. This is attributed to the high energy density deposition during the short duration of laser heating pulses. In the case of end of cooling period of the 10th pulse ($t = 20$ ms) temperature reduces considerably. The high rate of temperature decay is associated with the conduction, convection and radiation heat losses from the laser irradiated region. Since the temperature gradient is high in the heated region, conduction heat transfer from heated region to solid bulk is considerable while contributing to sharp temperature decay in the cooling period.

Figure 4.10 shows temperature distribution along the depth of the irradiated region for the laser intensity parameter $\beta = 1$. The behavior of temperature curves are similar to those shown in Fig. 4.9, provided that peak temperature and the temperature decay along the depth are not the same. This is attributed to the laser peak intensity, which moves away from the irradiated spot center for the laser intensity parameter $\beta = 1$. The differences in the behavior of temperature curves due to different laser intensity parameter are more pronounced during the heating cycle of the pulse. Since internal energy gain of the substrate material and temperature rise in the surface vicinity is associated with absorption of the incident laser energy, change in laser intensity distribution results in large change in the peak temperature at the coating surface.

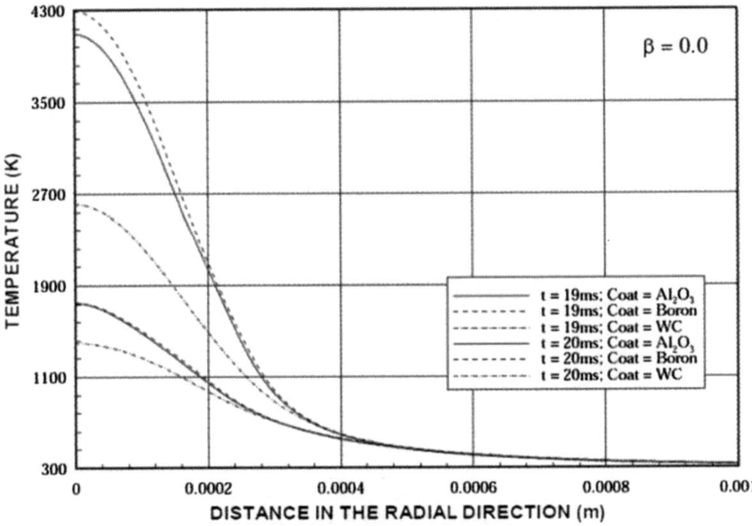

Fig. 4.11 Temperature distribution inside the substrate material along the radial distance for laser pulse intensity parameter $\beta = 0$, two heating periods and different coating materials

Figure 4.11 shows temperature distribution along the radial direction for different time periods, three materials, and laser intensity parameter $\beta = 0$. Temperature decays gradually in the central region of the irradiated spot around the symmetry axis where temperature is considerably high. As the distance in the radial direction increases away from the symmetry axis, temperature decay becomes sharp. This is attributed to the laser pulse intensity distribution at the workpiece surface, which is Gaussian. This is more pronounced at the end of heating period of 10th pulse (t = 19 ms) for tungsten carbide and alumina coatings. In the case of Tungsten Carbide, due to relatively high thermal diffusivity and low absorption coefficient, temperature becomes lower at the surface as compared to other coating materials. In the case of end of the cooling period (t = 20 ms), temperature drops significantly at the surface for all coating materials and temperature difference becomes small. This is attributed to the internal energy gain of the coating from the irradiated field, which ceases at the end of the heating period. Since laser beam power is switched off in the cooling cycle, heat losses due to convection, radiation and conduction from the surface lower temperature significantly at the surface region.

Figure 4.12 shows surface temperature distribution in the radial direction for the laser intensity parameter $\beta = 1$. Temperature distribution at the surface follows almost the laser pulse intensity distribution during the heating cycle (t = 9 ms). This is attributed to the material response to the incident laser radiation. The peak temperature exceeds the melting temperature of the coating material for alumina and boron; therefore, melting in the surface region is evident. However, no evidence of mushy zone is observed from the temperature profiles. This indicates that

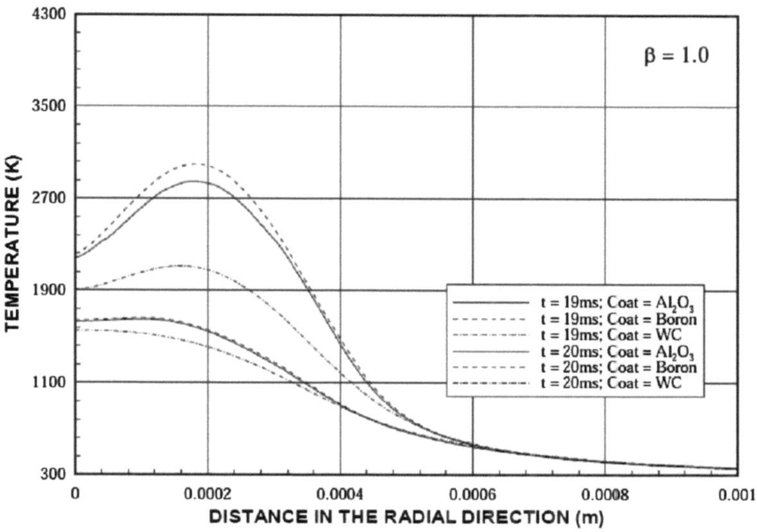

Fig. 4.12 Temperature distribution inside the substrate material along the radial distance for laser pulse intensity parameter $\beta = 1$, two heating periods and different coating materials

the small size of mushy zone is present between the liquid and solid phases, which may not be seen clearly from temperature curves. In the cooling cycle, absorption of incident radiation ceases so that temperature drops significantly at the surface. In this case, the peak temperature remains below the melting temperature of the substrate material.

Figures 4.13 and 4.14 show temporal variation of surface temperature for different coating materials for $\beta = 0$ and $\beta = 1$, respectively. Temperature increases first and reduces later during each pulse, which appears as oscillation in temperature profile with progressing time. Since the duty cycle is 50 %, the heating cycle is 1 ms while the cooling cycle ends after 1 ms of the heating cycle of the pulse. Therefore, the total heating and cooling durations of each pulse is 2 ms. Temperature rises sharply during the early period of heating cycle and temperature rise becomes gradual as the time progresses towards the end of heating cycle. This is attributed to absorption of incident beam in the early heating period; in which case, internal energy increase becomes rapid, which produces the rise of surface temperature at the same rate. As the heating period progresses, heat conduction from the irradiated region to the solid bulk increases because of high temperature gradient developed during the initial period of the heating cycle. In the case of cooling cycle, temperature reduces sharply due to cease of the laser power. However, the slope of temperature decay changes slightly towards the end of the cooling cycle. This change is associated with the phase change in the irradiated center; in which case, the slope of temperature decay changes in the vicinity of the melting temperature of the coating material. In the case of laser intensity parameter $\beta = 1$, temperature rise is considerably smaller than that of $\beta = 0$, since

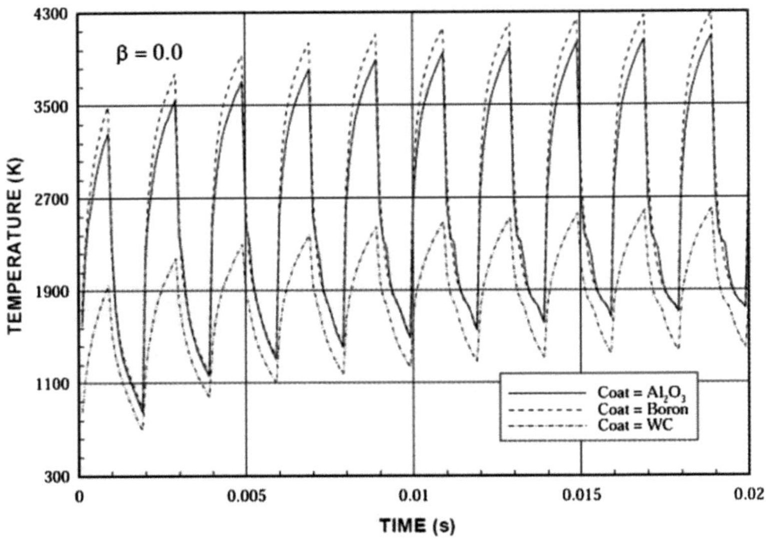

Fig. 4.13 Temporal variation of temperature at the irradiated spot center for laser pulse intensity parameter $\beta = 0$ and different coating materials

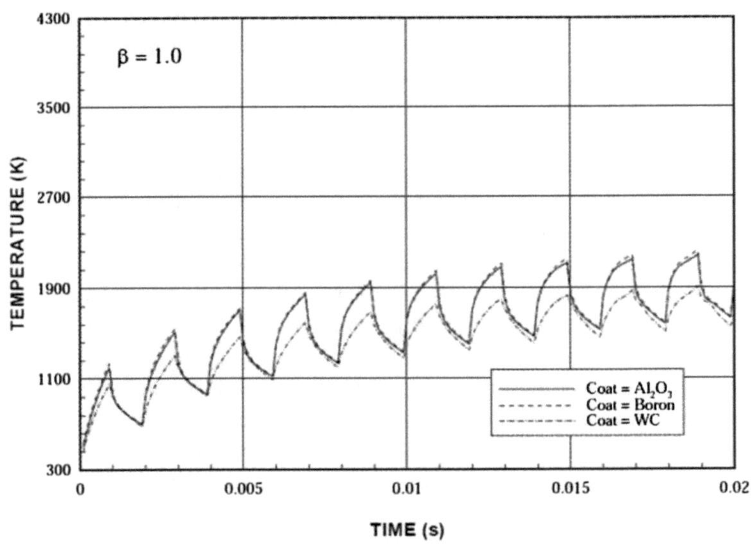

Fig. 4.14 Temporal variation of temperature at the irradiated spot center for laser pulse intensity parameter $\beta = 1$ and different coating materials

the peak intensity shifts away from the irradiated spot center for the laser intensity parameter $\beta = 1$. Surface temperature difference at the irradiated spot center becomes smaller for $\beta = 1$ than that of $\beta = 0$.

4.3.3 *Influence of Coating Material Thickness on Temperature and Flow Field in the Melt Pool*

Laser melting of tungsten carbide coated steel surfaces is considered to examine the influence of coating thickness on temperature field and melt zone in the laser heated region. The findings are discussed in the line with the previous study [3]. The properties used in the simulations for the materials and the laser pulse are given in Tables 4.2 and 4.4.

Figure 4.15 shows temperature distribution inside coating and steel for three coating thicknesses during the heating and cooling cycles of the 10th pulse and laser pulse parameter $\beta = 0$, which corresponds to a Gaussian intensity distribution at the irradiated surface. It should be noted that the total number of laser pulses is 10 during the heating process and more than 10 pulses, surface evaporation can occur, which is excluded from the analysis. Temperature decay is gradual in the surface region for all coating thicknesses considered. This is attributed to one or all of the followings: (1) energy gain from the irradiated field in the surface region is high, which in turn increases the internal energy in this region, (2) tungsten coating has low thermal diffusivity than steel which lowers heat diffusion from surface region steel sheet, and (3) phase change in the surface region suppresses heat diffusion from the melted zone to the solid bulk due to the presence of the mushy zone. It should be noted that temperature remains at solidus and liquidus temperatures across the mushy zone. Temperature decays sharply as

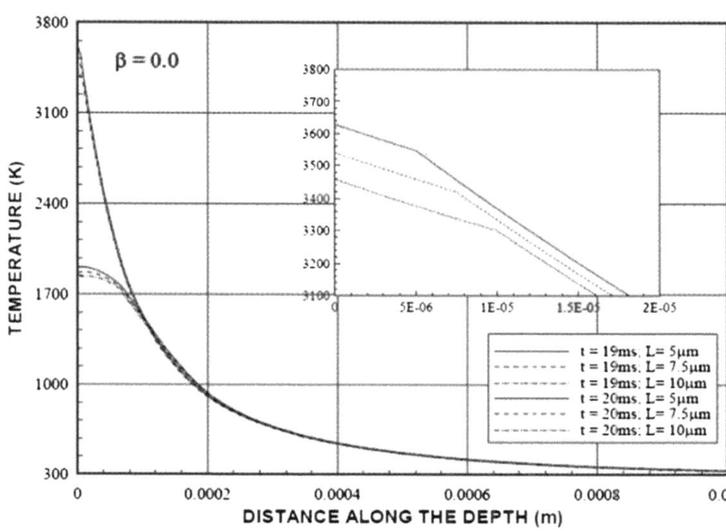

Fig. 4.15 Temperature distribution in the coating and steel for two different heating periods and different coating thicknesses (L) for the laser pulse parameter $\beta = 0$. t = 19 ms represents the end of heating cycle of 10th pulse and t = 20 ms corresponds to end of cooling cycle of 10th pulse

the distance increases below the surface, which is more pronounced in the region below the melting zone. The sharp increase in temperature is associated with heat conduction below the melt zone, which increases with increasing temperature gradient. The influence of coating thickness on the peak temperature at the surface (x = 0) is notable; in which case, small coating thickness results in high peak temperature. This is attributed to the thermal properties of tungsten carbide coating which has low specific heat capacity than steel; therefore, increasing coating thickness enhances temperature rise. This situation is less notable for the cooling period. It should be noted that the cooling period initiates immediately after ending of the consecutive pulse. In addition, the influence of coating thickness on temperature decay differs in the cooling and the heating cycles. In this case, temperature decay is gradually in the surface region as well as in the region next to the melt pool developed during the heating pulse. This is attributed to energy deposited, by the laser beam, to the substrate material in the surface region. Once the cooling cycle initiates, the internal energy gain from the irradiated field ceases in the surface region and heat transfer from the surface region though convection and conduction lowers temperature in this region. Figure 4.16 shows temperature distribution inside the substrate material for the laser pulse parameter $\beta = 1$. Temperature variation along the depth is similar to that corresponding to $\beta = 0$, provided that the peak temperature at the surface is lower than that of $\beta = 0$. This is because of the laser intensity distribution at the surface; in which case, the pulse parameter $\beta = 0$ results in high peak intensity at the irradiated spot center than

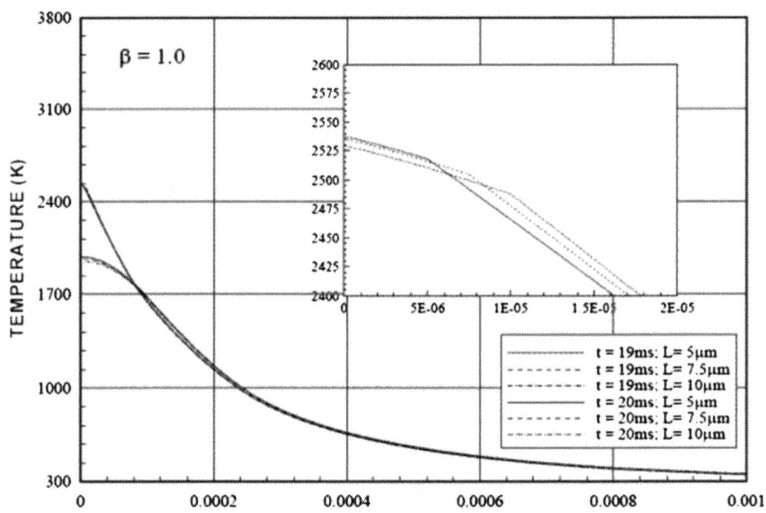

Fig. 4.16 Temperature distribution in the coating and steel for two different heating periods and different coating thicknesses (L) for the laser pulse parameter $\beta = 1$. t = 19 ms represents the end of heating cycle of 10th pulse and t = 20 ms corresponds to end of cooling cycle of 10th pulse

that corresponding to $\beta = 1$. In addition, the decay rate of temperature is gradual for $\beta = 1$ than that corresponding to $\beta = 0$. This is true for both heating and cooling periods and for all coating thicknesses. Consequently, influence of power intensity distribution on temperature in the surface region is significant while influence of coating thickness on temperature distribution in this region is small. Although increasing coating thickness lowers the peak temperature at the surface, temperature decay at the surface vicinity differs for $\beta = 1$ than that corresponding to $\beta = 0$. This is attributed to the phase change at the surface, which results in shallow melt depth at the coating surface vicinity while modifying temperature distribution in this region.

Figure 4.17 shows temperature distribution in the radial direction for three coating thicknesses, heating and cooling periods of 10th pulse, and the laser pulse parameter $\beta = 0$. Temperature decay across the irradiated spot center is gradual and as the distance in the radial direction increases further away from the irradiated spot center, temperature decay becomes sharp. Temperature profile follows almost the laser pulse intensity distribution along the radial direction. However, in the region of the melting (close to the symmetry axis where r = 0), temperature decay is considerably small because of energy stored in the substrate material during the phase change process. The peak temperature increases with reducing coating thickness. This is particularly true in the region close to the irradiated spot center where laser power intensity is high. As the distance increases away from the irradiated center, the differences in surface temperature due to the different coating

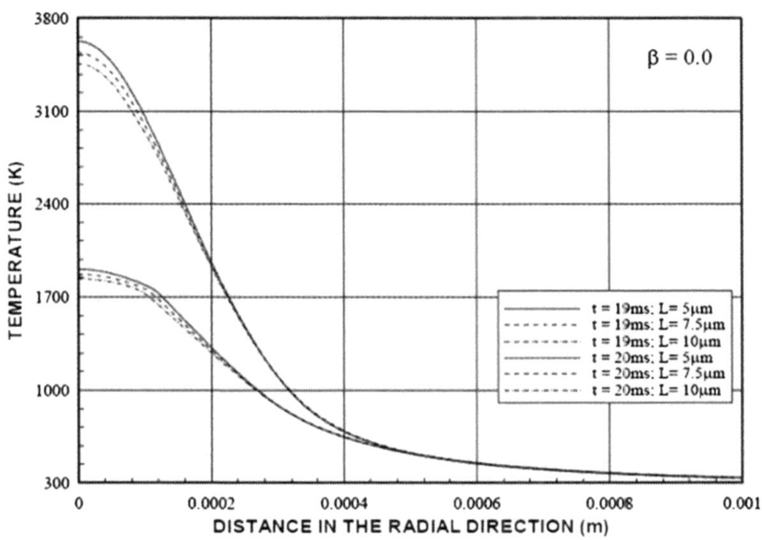

Fig. 4.17 Temperature distribution along the radial direction at the coating surface for two different heating periods, different coating thicknesses, and the laser pulse parameter $\beta = 0$. t = 19 ms represents the end of heating cycle of 10th pulse and t = 20 ms corresponds to end of cooling cycle of 10th pulse

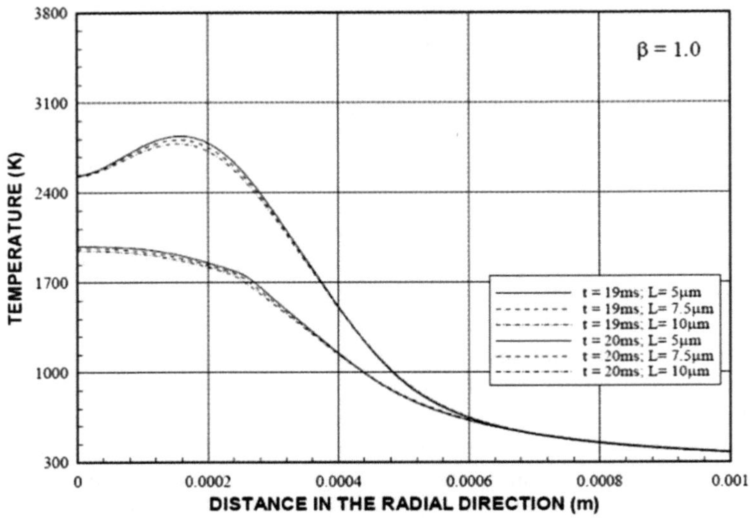

Fig. 4.18 Temperature distribution along the radial direction at the coating surface for two different heating periods, different coating thicknesses, and the laser pulse parameter $\beta = 1$. t = 19 ms represents the end of heating cycle of 10th pulse and t = 20 ms corresponds to end of cooling cycle of 10th pulse

thicknesses become negligibly small. In the case of end of the cooling cycle, temperature remains almost the same across the irradiated spot for all thicknesses considered. The peak temperature differences due to coating thickness become smaller as compared to that corresponding to the end of the cooling cycle. In addition, the peak temperature reduces significantly at the end of the cooling cycle, which is attributed to the internal energy loss from the irradiated region to its neighborhood through convection and conduction heat transfer. As the distance increases away from the irradiated region, temperature decay becomes relatively gradual as compared to that corresponding to end of the heating period. Figure 4.18 shows temperature distribution in the radial direction at the workpiece surface for the laser pulse parameter $\beta = 1$ and three coating thicknesses. The location of the peak temperature moves away from the irradiated spot center. The maximum surface temperature due to different coating thicknesses is less than those corresponding to $\beta = 0$. This, is associated with the laser peak power intensity, which is lower for $\beta = 1$ as compared with $\beta = 0$. In the case of the end of the cooling cycle, temperature remains almost the same along the radial direction covering the larger distance as compared to that of $\beta = 0$. In addition, temperature decay in the region away from the irradiated spot center is more gradual than that corresponding to $\beta = 0$. This indicates that influence of laser power intensity profile on surface temperature distribution is significant despite the fact that the laser pulse energy remains the same for two laser intensity parameters.

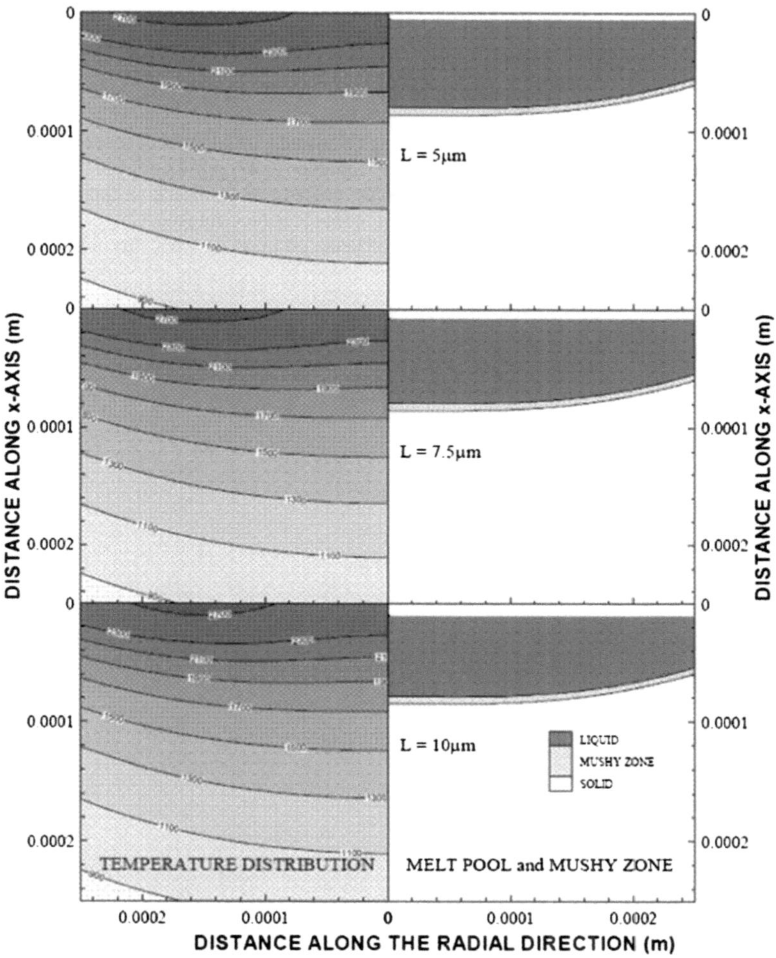

Fig. 4.19 Temperature contours, and melt depth and mushy zone inside the coating and steel for different coating thicknesses and the laser pulse parameter $\beta = 1$

Figure 4.19 show temperature contours and molten layer formed inside the substrate material for $\beta = 1$ and different coating thicknesses. It is evident that no melting is resulted in the coating while the melt pool is formed in steel below the coating despite the coating remains in the solid phase. This indicates that conduction heating from the coating results in phase change in steel substrate. However, the depth of the melt pool in steel reduces with increasing coating thickness. The presence of solid phase of coating is attributed to the lower beam intensity, which is lower for $\beta = 1$ than that of $\beta = 0$.

References

1. Shuja, S.Z., Yilbas, B.S.: Laser melting of alumina coated steel. AIChE J. **57**(9), 2547–2554 (2011)
2. Shuja, S.Z., Yilbas, B.S.: Laser induced heating of coated carbon steel sheets: consideration of melting and Marangoni flow, optics and laser technology. **47**, pp. 47–55 (2013)
3. Shuja, S.Z.: Laser heating of tungsten carbide coated steel surface: influence of coating thickness on temperature field and melt depth, numerical heat transfer (2012, in press)
4. Incropera, F.P., DeWitt, D.P.: Fundamentals of heat and mass transfer, 4th (ed.). Wiley, New York, pp. 829–830 (1996)

Chapter 5
Laser Induced Evaporation at the Surface

Abstract High power laser beam finds wide applications in industry including drilling, welding, cutting, and surface ablation. The common physical process involved with these applications is the phase change that takes place during the heating process and it is complicated due to rapid interactions during high power heating pulse. As the beam power increases, surface evaporation dominates over the melting, in which case ablation governs the heating process.

5.1 Introduction

High power laser beam finds wide applications in industry including drilling, welding, cutting, and surface ablation. The common physical process involved with these applications is the phase change that takes place during the heating process and it is complicated due to rapid interactions during high power heating pulse. As the beam power increases, surface evaporation dominates over the melting, in which case ablation governs the heating process. In this case, the heated surface temperature rises substantially and particle, thermionic, and vapor emissions take place. The thermionic emission is negligible at temperatures around the boiling temperature of the substrate material as compared to vapor emission. Moreover, as the heating progresses further, plasma is formed above the surface of the substrate material, which is irradiated by a laser beam. The plasma effect is negligible for laser machining intensities ($10^{10} < I_o < 10^{12}\ \mathrm{W/m^2}$, where I_o is the peak power intensity). Consequently, evaporation can be considered as the governing mechanism in laser machining process. In addition, for short pulse intensities, the nucleation formation and liquid expulsion from the irradiated zone is negligibly small; therefore, as the pulse intensity increases while the pulse length reduces, the phase change is governed by the evaporation process. In this chapter, the phase change due to evaporation at the surface is considered and the melting of the substrate material, plasma formation, and liquid ejection from the surface are

B. S. Yilbas and S. Z. Shuja, *Laser Surface Processing and Model Studies*,
Materials Forming, Machining and Tribology, DOI: 10.1007/978-3-642-36629-1_5,

neglected in the analysis. The evaporation at the surface is treated analytically and numerically in the light of the previous studies [1, 2].

5.2 Analytical Treatment of Evaporation

The energy equation for one dimensional solid is considered and the recession of the surface due to evaporation is incorporated in the analysis. The governing equation of heat transfer is solved analytically using a Laplace transform method. The closed form solution derived from the present study is validated through comparison with the previously obtained analytical solution for a conduction limited heating case [3]. The analysis presented herein is in line with the previous study [1].

The Fourier heating model governing unsteady heating, and allowing a phase change process, can be written as:

$$k\frac{\partial^2 T}{\partial x^2} + \rho CpV\frac{\partial T}{\partial x} + I_o(1 - r_f)\exp(-\beta t)\delta\exp(-\delta x) = \rho Cp\frac{\partial T}{\partial t} \tag{5.1}$$

where k is the thermal conductivity, ρ is the density, β is the pulse parameter, δ is the absorption coefficient, Cp is the specific heat capacity, and V is the recession velocity. The initial condition is:

$$\text{At time } t = 0 \rightarrow T(x, 0) = 0$$

The boundary conditions:

$$\text{At the surface } x = 0 \quad \rightarrow \left.\frac{\partial T}{\partial x}\right|_{x=0} = \frac{\rho V L_{ev}}{k}$$

$$x \text{ at infinity } x = \infty \rightarrow T(\infty, t) = 0$$

where V is the recession velocity, L_{ev} is the latent heat of evaporation and r_f is the surface reflectivity. The recession velocity of the surface is similar to the surface melting case (Eq. 3.5), provided that L_m is replaced with L_{ev}. Therefore, recession velocity due to surface evaporation yields:

$$V = \frac{I_1}{\rho[CpTs + L_{ev}]} \tag{5.2}$$

where $I_1 = I_o(1 - r_f)$. Equation 5.1 can be written as:

$$\frac{\partial^2 T}{\partial x^2} + \frac{V}{\alpha}\frac{\partial T}{\partial x} + \frac{I_1}{k}\exp(-\beta t)\delta\exp(-\delta x) = \frac{1}{\alpha}\frac{\partial T}{\partial t} \tag{5.3}$$

The Laplace transform of Eq. 5.3 with respect to t can be written as:

$$\frac{\partial^2 \overline{T}}{\partial x^2} + \frac{V}{\alpha}\frac{\partial \overline{T}}{\partial x} + I_1 \frac{\delta}{k}\frac{1}{(p+\beta)}\exp(-\delta x) = \frac{1}{\alpha}\left[p\overline{T}\right] \quad (5.4)$$

where $\overline{T}$ is the Laplace transform of the temperature. Equation 5.4 yields:

$$\frac{\partial^2 \overline{T}}{\partial x^2} + \frac{V}{\alpha}\frac{\partial \overline{T}}{\partial x} - \frac{p}{\alpha}\overline{T} = -I_1 \frac{\delta}{k}\frac{1}{(p+\beta)}\exp(-\delta x) \quad (5.5)$$

Equation 5.5 has homogeneous and particular solutions. Therefore, $\overline{T}$ yields:

$$\overline{T} = \overline{Th} + \overline{Tp} \quad (5.6)$$

The characteristic equation for the homogeneous solution can be written as:

$$r^2 + \frac{V}{\alpha}r - \frac{p}{\alpha} = 0 \quad (5.7)$$

It yields the solution of:

$$r_{1,2} = -\frac{V}{2\alpha} \pm \frac{\sqrt{V^2 + 4p\alpha}}{2\alpha} \quad (5.8)$$

Therefore, the homogenous solution $\left(\overline{Th}\right)$ yields:

$$\overline{Th} = c_1 e^{r_1 x} + c_2 e^{r_2 x} \quad (5.9)$$

or,

$$\overline{Th} = e^{-\frac{Vx}{2\alpha}}\left[c_1 e^{-\frac{\sqrt{V^2+4p\alpha}}{2\alpha}x} + c_2 e^{\frac{\sqrt{V^2+4p\alpha}}{2\alpha}x}\right] \quad (5.10)$$

For the particular solution, one can propose $\overline{Tp} = A_o e^{-\delta x}$. Consequently, substitution of $\overline{Tp}$ in Eq. 5.5 results:

$$A_o\delta^2 e^{-\delta x} - \frac{V}{\alpha}\delta A_o e^{-\delta x} - \frac{p}{\alpha}A_o e^{-\delta x} = H_o e^{-\delta x} \quad (5.11)$$

where

$$H_o = -I_1 \frac{\delta}{k}\frac{1}{(p+\beta)} \quad (5.12)$$

and

$$A_o = -\frac{H_o\alpha}{p + V\delta - \alpha\delta^2} \quad (5.13)$$

or

$$\overline{T} = e^{-\frac{Vx}{2\alpha}}\left[c_1 e^{-\frac{\sqrt{V^2+4p\alpha}}{2\alpha}x} + c_2 e^{\frac{\sqrt{V^2+4p\alpha}}{2\alpha}x}\right] - \frac{H_o\alpha}{p + V_1}e^{-\delta x} \tag{5.14}$$

where $V_1 = V\delta - \alpha\delta^2$.

Since $T(\infty, t) = 0$, then $c_2 = 0$.

Therefore:

$$\overline{T} = c_1 e^{-\frac{1}{2\alpha}\left[V+\sqrt{V^2+4\alpha p}\right]x} + \frac{I_1\delta\alpha}{k}\frac{e^{-\delta x}}{(p+\beta)(p+V_1)} \tag{5.15}$$

Let $w_1 = \frac{I_1\delta\alpha}{k}$ and use the boundary condition $\left[\frac{\partial T}{\partial x}\right]_{x=0} = \frac{\rho VL}{k}$, one can obtain c_1, i.e.:

$$\frac{\partial\overline{T}}{\partial x} = \left\{\begin{array}{l} -\dfrac{1}{2\alpha}\left[V+\sqrt{V^2+4\alpha p}\right]c_1 e^{-\frac{V}{2\alpha}\left[V+\sqrt{V^2+4\alpha p}\right]x} \\ -\dfrac{\delta w_1 e^{-\delta x}}{(p+\beta)(p+V_1)} \end{array}\right\}_{x=0} = \frac{\rho VL_{ev}}{kp} \tag{5.16}$$

or

$$c_1 = -\frac{2\alpha\delta w_1}{(p+\beta)(p+V_1)\left[V+\sqrt{V^2+4\alpha p}\right]} - \frac{\frac{\rho VL}{k}2\alpha}{\left(V+\sqrt{V^2+4\alpha p}\right)p} \tag{5.17}$$

Hence:

$$\overline{T} = -\frac{2\alpha\delta w_1 e^{-\frac{1}{2\alpha}\left[V+\sqrt{V^2+4\alpha p}\right]x}}{(p+\beta)(p+V_1)\left[V+\sqrt{V^2+4\alpha p}\right]} + \frac{w_1 e^{-\delta x}}{(p+\beta)(p+V_1)} + \overline{T_{mo}} \tag{5.18}$$

where

$$\overline{T_{mo}} = -2\alpha\frac{\rho VL}{k}\frac{e^{-\frac{1}{2\alpha}\left[V+\sqrt{V^2+4\alpha p}\right]x}}{p\left(V+\sqrt{V^2+4\alpha p}\right)} \tag{5.19}$$

Let H_1, H_2, and H_5 be:

$$H_1 = \frac{e^{-\frac{1}{2\alpha}\left[V+\sqrt{V^2+4\alpha p}\right]x}}{(p+\beta)(p+V)_1\left[V+\sqrt{V^2+4\alpha p}\right]} \tag{5.20}$$

$$H_2 = \frac{1}{(p+\beta)(p+V_1)} = \frac{1}{(V_1-\beta)}\left[\frac{1}{p+\beta} - \frac{1}{p+V_1}\right] \tag{5.21}$$

and

$$H_5 = \frac{e^{-\frac{1}{2\alpha}[V + \sqrt{V^2 + 4\alpha p}]x}}{p\left(V + \sqrt{V^2 + 4\alpha p}\right)} \tag{5.22}$$

Hence,

$$T(x,\,t) = L^{-1}\overline{T} = -2\alpha\delta w_1 L^{-1}H_1 + w_1 e^{-\delta x}L^{-1}H_2 - 2\alpha\frac{\rho VL}{k}L^{-1}H_5 \tag{5.23}$$

To obtain the inverse transformation of functions H_1 and H_5, let introduce:

$$s = V^2 + 4\alpha p \,\text{or}\, ds = 4\alpha dp \,\text{and}\, p = \frac{1}{4\alpha}\left(s - V^2\right) \tag{5.24}$$

Therefore:

$$L^{-1}H_1 = \frac{1}{2\pi i}\int_{c - i\infty}^{c + i\infty} e^{tp}H_1(P)dp \tag{5.25}$$

or

$$L^{-1}H_1 = \frac{1}{4\alpha}e^{-\left(\frac{Vx}{2\alpha} + \frac{tV^2}{4\alpha}\right)}\left\{\frac{1}{2\pi i}\int_{\bar{c} - i\infty}^{\bar{c} + i\infty}\frac{e^{-\left(\frac{x\sqrt{s}}{2\alpha} - \frac{st}{4\alpha}\right)}}{\left(\frac{s}{4\alpha} - \frac{V^2}{4\alpha} + \beta\right)\left(\frac{s}{4\alpha}\frac{V^2}{4\alpha} + V\right)\left(V + \sqrt{s}\right)}ds\right\} \tag{5.26}$$

where $\bar{c} = 4\alpha c + V^2$.

Employing one more transformation by letting:

$$\sigma = \frac{s}{4\alpha} \text{ or } ds = 4\alpha d\sigma \tag{5.27}$$

and, after algebraic manipulation, the following was obtained:

$$L^{-1}H_1 = e^{-\frac{V}{2\alpha}\left(x + \frac{V}{2}t\right)}\left\{\frac{1}{2\pi i}\int_{\widehat{c} - i\infty}^{\widehat{c} + i\infty}\frac{e^{-\frac{x\sqrt{p}}{\sqrt{\alpha}}}e^{\sigma t}}{\left(\sigma - \frac{V^2}{4\alpha} + \beta\right)\left(\sigma - \frac{V^2}{4\alpha} + V_1\right)\left(V + 2\sqrt{\alpha}\sqrt{\sigma}\right)}d\sigma\right\} \tag{5.28}$$

where $\widehat{c} = \frac{\bar{c}}{4\alpha}$.

Hence:

$$L^{-1}H_1 = e^{-\frac{V}{2\alpha}\left(x + \frac{V}{2}t\right)}L^{-1}H_3 \tag{5.29}$$

where:

$$H_3 = \frac{e^{-\left(\frac{x}{\sqrt{\alpha}}\sqrt{p}\right)}}{(p - \frac{V^2}{4\alpha} + \beta)(p - \frac{V^2}{4\alpha} + V_1)\left(V + 2\sqrt{\alpha}\sqrt{p}\right)} \tag{5.30}$$

Similarly, $L^{-1}H_5$ can be obtained, i.e.:

$$L^{-1}H_5 = e^{-\frac{V}{2\alpha}\left(x + t\frac{V}{2}\right)}\left\{\frac{1}{2\pi i}\int_{c - i\infty}^{c + i\infty} \frac{e^{-\frac{x\sqrt{\sigma}}{\sqrt{\alpha}}}e^{\sigma t}}{(V + 2\sqrt{\alpha}\sqrt{\sigma})}d\sigma\right\} = 4\alpha e^{-\frac{V}{2\alpha}\left(x + t\frac{V}{2}\right)}L^{-1}H_7 \tag{5.31}$$

where:

$$H_7 = \frac{e^{-\left(\frac{x}{\sqrt{\alpha}}\sqrt{p}\right)}}{\left(\sqrt{\alpha}\sqrt{p} - V2\right)\left(2\sqrt{\alpha}\sqrt{p} + V\right)^2} \tag{5.32}$$

or

$$H_7 = \frac{1}{8\alpha\sqrt{\alpha}}\frac{e^{-\left(\frac{x}{\sqrt{\alpha}}\sqrt{p}\right)}}{\left(\sqrt{p} - \eta\right)\left(\sqrt{p} + \eta^2\right)} \tag{5.33}$$

where $\eta = \frac{V}{2\sqrt{\alpha}}$.

Introducing partial fractions, re-arrangement produces:

$$H_7 = \frac{1}{8\sqrt{\alpha}V^2}\left[\frac{e^{-\left(\frac{x}{\sqrt{\alpha}}\sqrt{p}\right)}}{\sqrt{p} - \eta} - \frac{e^{-\left(\frac{x}{\sqrt{\alpha}}\sqrt{p}\right)}}{\left(\sqrt{p} + \eta\right)} - \frac{V}{\sqrt{\alpha}}\frac{e^{-\left(\frac{x}{\sqrt{\alpha}}\sqrt{p}\right)}}{\left(\sqrt{p} + \eta\right)^2}\right] \tag{5.34}$$

It is noted from the Laplace inversion that [4]:

$$L^{-1}\frac{e^{-k\sqrt{p}}}{a + \sqrt{p}} = \frac{1}{\sqrt{\pi t}}\exp\left(-\frac{k^2}{4t}\right) - ae^{ak}e^{a^2 t}\mathrm{erfc}\left(a\sqrt{t} + \frac{k}{2\sqrt{t}}\right) \tag{5.35}$$

where $k \geq 0$. Therefore:

$$L^{-1}\frac{e^{-\left(\frac{x}{\sqrt{\alpha}}\sqrt{p}\right)}}{\sqrt{p} - \eta} = \frac{1}{\sqrt{\pi t}}e^{\left(-\frac{x^2}{4\alpha t}\right)} + \frac{V}{2\sqrt{\alpha}}e^{-\frac{Vx}{2\alpha}}e^{\frac{V^2}{4\alpha}t}\mathrm{erfc}\left(\frac{x}{2\sqrt{\alpha t}} - \frac{V}{2\sqrt{\alpha}}\sqrt{t}\right) \tag{5.36}$$

Similarly,

$$L^{-1}\frac{e^{-\left(\frac{x}{\sqrt{\alpha}}\sqrt{p}\right)}}{\sqrt{p} + \eta} = \frac{1}{\sqrt{\pi t}}e^{\left(-\frac{x^2}{4\alpha t}\right)} - \frac{V}{2\sqrt{\alpha}}e^{\frac{Vx}{2\alpha} + \frac{V^2 t}{4\alpha}}\mathrm{erfc}\left(\frac{x}{2\sqrt{\alpha t}} + \frac{V}{2\sqrt{\alpha}}\sqrt{t}\right) \tag{5.37}$$

Letting $\sqrt{p} + \eta = \sqrt{s}$ and using the definition of inverse Laplace integral $L^{-1}\frac{e^{-\left(\frac{x}{\sqrt{\alpha}}\sqrt{p}\right)}}{\left(\sqrt{p} + \eta\right)^2}$ becomes:

$$L^{-1}\frac{e^{-\left(\frac{x}{\sqrt{\alpha}}\sqrt{p}\right)}}{\left(\sqrt{p}+\eta\right)^2}=e^{\frac{V^2t}{4\alpha}+\frac{Vx}{2\alpha}}\left(\operatorname{erfc}\left(\frac{V\sqrt{t}}{2\sqrt{\alpha}}+\frac{x}{2\sqrt{\alpha t}}\right)\left(1+\frac{Vx}{2\alpha}+\frac{V^2t}{2\alpha}\right)\right)-\frac{V\sqrt{t}}{\sqrt{\pi\alpha}}e^{\left(-\frac{x^2}{4\alpha t}\right)} \tag{5.38}$$

After simplifications, the Laplace inversion of H_5 becomes:

$$L^{-1}H_5=\frac{1}{16\alpha V}\left(\begin{array}{l}e^{-\frac{Vx}{\alpha}}\operatorname{erfc}\left(\frac{x-Vt}{2\sqrt{\alpha t}}\right)\\-\left(1+\frac{Vx}{\alpha}+\frac{tV^2}{\alpha}\right)\operatorname{erfc}\left(\frac{x+Vt}{2\sqrt{\alpha t}}\right)\end{array}\right)+\frac{\sqrt{t}}{8\alpha\sqrt{\pi\alpha}}e^{-\frac{(x+Vt)^2}{4\alpha t}} \tag{5.39}$$

Let:

$$w_2^2=\frac{V^2}{4\alpha}-\beta\ :\ k_1=\frac{x}{\sqrt{\alpha}}\ :\ w_3^2=\frac{V^2}{4\alpha}-V_1\ :\ w_4=\frac{V}{2\sqrt{\alpha}} \tag{5.40}$$

then, H_3 becomes:

$$H_3=\frac{1}{\left[\left(\sqrt{p}\right)^2-w_2^2\right]\left[\left(\sqrt{p}\right)^2-w_3^2\right]\left[\sqrt{p}+w_4\right]}\frac{1}{2\sqrt{\alpha}}e^{-k_1\sqrt{p}} \tag{5.41}$$

and after using partial fraction expansion, H_3 becomes:

$$H_3=\left[\frac{D_1}{\sqrt{p}-w_2}+\frac{D_2}{\sqrt{p}+w_2}+\frac{D_3}{\sqrt{p}-w_3}+\frac{D_4}{\sqrt{p}+w_3}+\frac{D_5}{\sqrt{p}+w_4}\right]\frac{1}{2\sqrt{\alpha}}e^{-k_1\sqrt{p}} \tag{5.42}$$

where:

$$\begin{aligned}D_1&=\frac{1}{2w_2\left(w_2^2-w_3^2\right)\left(w_2+w_4\right)}\\D_2&=-\frac{1}{2w_2\left(w_2^2-w_3^2\right)\left(-w_2+w_4\right)}\\D_3&=\frac{1}{2w_3\left(w_3^2-w_2^2\right)\left(w_3+w_4\right)}\\D_4&=-\frac{1}{2w_3\left(w_3^2-w_2^2\right)\left(-w_3+w_4\right)}\\D_5&=\frac{1}{\left(w_4^2-w_2^2\right)\left(w_4^2-w_3^2\right)}\end{aligned} \tag{5.43}$$

$L^{-1}H_3$ can be written as:

$$L^{-1}H_3 = \frac{1}{2\sqrt{\alpha}}\left\{\begin{array}{l} D_1\left[\frac{1}{\sqrt{\pi t}}e^{-\frac{k_1^2}{4t}} + w_2 e^{-w_2 k_1} e^{w_2^2 t}\operatorname{erfc}\left(-w_2\sqrt{t} + \frac{k_1}{2\sqrt{t}}\right)\right] \\ +D_2\left[\frac{1}{\sqrt{\pi t}}e^{-\frac{k_1^2}{4t}} - w_2 e^{w_2 k_1} e^{w_2^2 t}\operatorname{erfc}\left(w_2\sqrt{t} + \frac{k_1}{2\sqrt{t}}\right)\right] \\ +D_3\left[\frac{1}{\sqrt{\pi t}}e^{-\frac{k_1^2}{4t}} + w_3 e^{-w_3 k_1} e^{w_3^2 t}\operatorname{erfc}\left(-w_3\sqrt{t} + \frac{k_1}{2\sqrt{t}}\right)\right] \\ +D_4\left[\frac{1}{\sqrt{\pi t}}e^{-\frac{k_1^2}{4t}} - w_3 e^{w_3 k_1} e^{w_3^2 t}\operatorname{erfc}\left(w_3\sqrt{t} + \frac{k_1}{2\sqrt{t}}\right)\right] \\ +D_5\left[\frac{1}{\sqrt{\pi t}}e^{-\frac{k_1^2}{4t}} - w_4 e^{w_4 k_1} e^{w_4^2 t}\operatorname{erfc}\left(w_4\sqrt{t} + \frac{k_1}{2\sqrt{t}}\right)\right] \end{array}\right\} \tag{5.44}$$

Since, $T(x, t) = L^{-1}\overline{T}$, therefore, it yields:

$$T(x, t) = L^{-1}\overline{T} = -2\alpha\delta w_1 e^{-\frac{V}{2\alpha}(x + \frac{V}{2}t)} L^{-1}H_3 + w_1 e^{-\delta x} L^{-1}H_2 - 8\alpha^2 \frac{\rho V L}{k} L^{-1}H_5 \tag{5.45}$$

Noting that:

$$L^{-1}H_2 = \frac{1}{V_1 - \beta}\left(e^{-\beta t} - e^{-V_1 t}\right) \tag{5.46}$$

Therefore, the Laplace transformation yields:

$$\begin{aligned} T(x, t) = -\frac{\alpha\delta w_1}{\sqrt{\alpha}} e^{-\frac{V}{2\alpha}(x + \frac{V}{2}t)} &\left\{\begin{array}{l} \frac{1}{\sqrt{\pi t}}e^{-\frac{k_1^2}{4t}}[D_1 + D_2 + D_3 + D_4 + D_5] \\ +D_1\left[w_2 e^{-w_2 k_1} e^{w_2^2 t}\operatorname{erfc}\left(-w_2\sqrt{t} + \frac{k_1}{2\sqrt{t}}\right)\right] \\ +D_2\left[-w_2 e^{-w_2 k_1} e^{w_2^2 t}\operatorname{erfc}\left(w_2\sqrt{t} + \frac{k_1}{2\sqrt{t}}\right)\right] \\ +D_3\left[w_3 e^{-w_3 k_1} e^{w_3^2 t}\operatorname{erfc}\left(-w_3\sqrt{t} + \frac{k_1}{2\sqrt{t}}\right)\right] \\ +D_4\left[-w_3 e^{-w_3 k_1} e^{w_3^2 t}\operatorname{erfc}\left(w_3\sqrt{t} + \frac{k_1}{2\sqrt{t}}\right)\right] \\ +D_5\left[-w_4 e^{-w_4 k_1} e^{w_4^2 t}\operatorname{erfc}\left(w_4\sqrt{t} + \frac{k_1}{2\sqrt{t}}\right)\right] \\ +\frac{w_1}{V_1 - \beta}e^{-\delta x}\left[e^{-\beta t} - e^{-V_1 t}\right] \end{array}\right\} \\ &- \frac{\alpha\rho L}{2k}\left(e^{-\frac{Vx}{\alpha}}\operatorname{erfc}\left(\frac{x - Vt}{2\sqrt{\alpha t}}\right) - \left(1 + \frac{Vx}{\alpha} + \frac{V^2 t}{\alpha}\right)\operatorname{erfc}\left(\frac{x + Vt}{2\sqrt{\alpha t}}\right)\right) \\ &- \frac{\rho V L}{k}\frac{\sqrt{\alpha t}}{\sqrt{\pi}}e^{-\frac{(x + Vt)^2}{4\alpha t}} \end{aligned} \tag{5.47}$$

However, $D_1 + D_2 + D_3 + D_4 + D_5 = 0$. Therefore, Eq. 5.47 becomes:

$$
\begin{aligned}
T(x,t) = & -\frac{I_1 \alpha \sqrt{\alpha}\delta^2}{k} e^{-\frac{V}{2\alpha}\left(x + \frac{Vt}{2}\right)} \\
& \times \left\{ \begin{array}{l}
\frac{e^{\left(\frac{V^2}{4\alpha}-\beta\right)t}}{2\left(V\delta - \alpha\delta^2 - \beta\right)} \left[e^{-\sqrt{\frac{V^2}{4\alpha}-\beta}\frac{x}{\sqrt{\alpha}}} \frac{\mathrm{erfc}\left(-\sqrt{\left(\frac{V^2}{4\alpha}-\beta\right)t} + \frac{x}{2\sqrt{\alpha t}}\right)}{\sqrt{\frac{V^2}{4\alpha}-\beta} + \frac{V}{2\sqrt{\alpha}}} + e^{\sqrt{\frac{V^2}{4\alpha}-\beta}\frac{x}{\sqrt{\alpha}}} \frac{\mathrm{erfc}\left(\sqrt{\left(\frac{V^2}{4\alpha}-\beta\right)t} + \frac{x}{2\sqrt{\alpha t}}\right)}{\frac{V}{2\sqrt{\alpha}} - \sqrt{\frac{V^2}{4\alpha}-\beta}} \right] \\
+ \frac{e^{\left(\frac{V^2}{4\alpha} - \left(V\delta - \alpha\delta^2\right)\right)t}}{2\left(\alpha\delta^2 - V\delta + \beta\right)} \left[e^{-\sqrt{\frac{V^2}{4\alpha} - \left(V\delta - \alpha\delta^2\right)}\frac{x}{\sqrt{\alpha}}} \frac{\mathrm{erfc}\left(-\sqrt{\left(\frac{V^2}{4\alpha} - \left(V\delta - \alpha\delta^2\right)\right)t} + \frac{x}{2\sqrt{\alpha t}}\right)}{\sqrt{\frac{V^2}{4\alpha} - \left(V\delta - \alpha\delta^2\right)} + \frac{V}{2\sqrt{\alpha}}} \right] \\
+ e^{\sqrt{\frac{V^2}{4\alpha} - \left(V\delta - \alpha\delta^2\right)}\frac{x}{\sqrt{\alpha}}} \frac{\mathrm{erfc}\left(\sqrt{\left(\frac{V^2}{4\alpha} - \left(V\delta - \alpha\delta^2\right)\right)t} + \frac{x}{2\sqrt{\alpha t}}\right)}{\frac{V}{2\sqrt{\alpha}} - \sqrt{\frac{V^2}{4\alpha} - \left(V\delta - \alpha\delta^2\right)}} \\
- \frac{V}{2\sqrt{\alpha}\beta\left(V\delta - \alpha\delta^2\right)} e^{\left(\frac{V^2}{4\alpha} - \beta\right)t} e^{\sqrt{\frac{V^2}{4\alpha}-\beta}\frac{x}{\sqrt{\alpha}}} \mathrm{erfc}\left(\sqrt{\left(\frac{V^2}{4\alpha} - \beta\right)t} + \frac{x}{2\sqrt{\alpha t}}\right)
\end{array} \right. \\
& + \frac{I_1 \delta \alpha}{k} \frac{1}{\left(V\delta - \alpha\delta^2\right) - \beta} e^{-\delta x}\left(e^{-\beta t} - e^{-\left(V\delta - \alpha\delta^2\right)t}\right) \\
& - \frac{\alpha \rho L}{2k}\left(e^{-\frac{Vx}{\alpha}} \mathrm{erfc}\left(\frac{x - Vt}{2\sqrt{\alpha t}}\right) - \left(1 + \frac{Vx}{\alpha} + \frac{V^2 t}{\alpha}\right)\mathrm{erfc}\left(\frac{x + Vt}{2\sqrt{\alpha t}}\right)\right) \\
& - \frac{\rho V L}{k}\frac{\sqrt{\alpha t}}{\sqrt{\pi}} e^{-\frac{(x+V)t^2}{4\alpha t}}
\end{aligned}
\tag{5.48}
$$

The surface temperature can be obtained after replacing $x = 0$ at the surface, then Eq. 5.48 reduces to the surface temperature, i.e.:

$$
\begin{aligned}
T(0,t) = & -\sqrt{\alpha}\delta w_1 e^{-\frac{V^2 t}{4\alpha}} \left\{ \begin{array}{c}
\frac{e^{w_2^2 t}}{(V_1 - \beta)(w_2 + w_4)}\left[1 + \frac{w_2}{w_4 - w_2}\mathrm{erfc}\left(w_2\sqrt{t}\right)\right] \\
-\frac{e^{w_3^2 t}}{(V_1 - \beta)(w_3 + w_4)}\left[1 + \frac{w_3}{w_4 - w_3}\mathrm{erfc}\left(w_3\sqrt{t}\right)\right] \\
-D_5 w_4 e^{w_4^2 t}\mathrm{erfc}\left(w_4\sqrt{t}\right)
\end{array} \right\} \\
& + \frac{w_1}{V_1 - \beta}\left(e^{-\beta t} - e^{-V_1 t}\right) \\
& - \frac{\alpha \rho L}{2k}\left(\mathrm{erfc}\left(-\frac{Vt}{2\sqrt{\alpha t}}\right) - \left(1 + \frac{V^2 t}{\alpha}\right)\mathrm{erfc}\left(\frac{Vt}{2\sqrt{\alpha t}}\right)\right) \\
& - \frac{\rho V L}{k}\frac{\sqrt{\alpha t}}{\sqrt{\pi}} e^{-\frac{(Vt)^2}{4\alpha t}}
\end{aligned}
\tag{5.49}
$$

It is known that:

$$
\alpha = \frac{k}{\rho C p} : V_1 = V\delta - \alpha\delta^2 : w_1 = \frac{I_1 \alpha \delta}{k} : w_2^2 = \frac{V^2}{4\alpha} - \beta
$$

$$
w_3^2 = \frac{V^2}{4\alpha} - V_1 : w_4^2 = \frac{V^2}{4\alpha} : k_1 = \frac{x}{\sqrt{\alpha}}
$$

$$
D_5 = \frac{1}{\left(w_4^2 - w_2^2\right)\left(w_4^2 - w_3^2\right)} = \frac{1}{\beta V_1}
$$

Substituting these values into Eq. 5.49 and after the algebraic manipulation yields:

$$T(0,t) = \frac{I_1 \alpha\delta}{k}\frac{1}{(V_1-\beta)}\left\{\begin{array}{c} \frac{-\delta\sqrt{\alpha}}{\beta} w_2 e^{-\beta t}\left[\mathrm{erfc}(w_2\sqrt{t})\right] \\ +\frac{\delta\sqrt{\alpha}}{V_1} w_3 e^{-V_1 t}\left[\mathrm{erfc}\left(\mathrm{w}_3\sqrt{\mathrm{t}}\right)\right] \\ +\frac{(V_1-\beta)}{\beta V_1}\delta\sqrt{\alpha} w_4 \mathrm{erfc}\left(w_4\sqrt{t}\right) \\ \left(\frac{-\delta\sqrt{\alpha}}{w_2+w_4}+1\right)e^{-\beta t}+\left(\frac{\delta\sqrt{\alpha}}{w_2+w_4}-1\right)e^{-V_1 t} \end{array}\right\}$$
$$-\frac{\alpha\rho L}{2k}\left(\mathrm{erfc}\left(-\frac{\mathrm{Vt}}{2\sqrt{\alpha \mathrm{t}}}\right)-\left(1+\frac{\mathrm{V}^2\mathrm{t}}{\alpha}\right)\mathrm{erfc}\left(\frac{\mathrm{Vt}}{2\sqrt{\alpha\mathrm{t}}}\right)\right)$$
$$-\frac{\rho VL}{k}\frac{\sqrt{\alpha t}}{\sqrt{\pi}}e^{-\frac{(Vt)^2}{4\alpha t}} \tag{5.50}$$

In order to obtain dimensionless Eq. 5.50, the followings are introduced:

$$V' = \frac{1}{\alpha\delta}V : \beta' = \frac{1}{\alpha\delta^2}\beta : V' = 1 + \frac{V_1}{\alpha\delta^2} \text{ and } \tau = \alpha\delta^2 t$$
$$w_2^2 = \left(\frac{V'}{4}-\beta'\right)\alpha\delta^2 : w_3^2 = \left(\frac{V'-2}{2}\right)^2\alpha\delta^2 : w_4^2 = \frac{V'}{2}\sqrt{\alpha}\delta$$
$$D_5 = \frac{1}{\beta V_1} = \frac{1}{\beta'(V'-1)(\alpha\delta^2)^2}$$

The surface temperature yields:

$$T(0,\tau) = \frac{I_1}{k\delta}\frac{1}{(V'-\beta'-1)}\left\{\begin{array}{c} \frac{-\sqrt{\frac{V'^2}{4}-\beta'}}{\beta'}e^{-\beta'\tau}\left[\mathrm{erfc}\left(\sqrt{\left(\frac{V'}{4}-\beta'\right)\tau}\right)\right] \\ +\frac{(V'-2)}{2(V'1)}e^{-(V'-1)\tau}\left[\mathrm{erfc}\left(\left(\frac{V'-2}{2}\right)\sqrt{\tau}\right)\right] \\ +\frac{(V'-\beta'-1)V'}{2\beta'(V'-1)}\mathrm{erfc}\left(\frac{V'}{2}\sqrt{\tau}\right) \\ \left(\frac{-1}{\sqrt{\frac{V'^2}{4}-\beta'}+\frac{V'}{2}}+1\right)e^{-\beta'\tau}+\frac{(V'-2)}{(V'-1)}e^{-(V'-1)\tau} \end{array}\right\}$$
$$-\frac{\alpha\rho L}{2k}\left[\left(\mathrm{erfc}\left(-\frac{V'\tau}{2\sqrt{\tau}}\right)-(1+V'^2\tau)\mathrm{erfc}\left(\frac{V'\tau}{2\sqrt{\tau}}\right)\right)+\frac{2V'\tau}{\sqrt{\pi}}e^{-\left(\frac{V'^2}{4\tau}+\frac{V'^2\tau}{4}\right)}\right] \tag{5.51}$$

Noting that:

$$1-\mathrm{erfc}(\xi) = \mathrm{erf}(\xi) \text{ and } \mathrm{erfc}(-\xi) = 2-\mathrm{erfc}(\xi) \tag{5.52}$$

Finally, the dimensionless surface temperature becomes:

$$\frac{T(0,\tau)}{\frac{I_1}{k\delta}} = \frac{1}{(V' - \beta' - 1)} \left\{ \begin{array}{c} \frac{-\sqrt{\frac{V'^2}{4} - \beta'}}{\beta'} e^{-\beta'\tau} \left[\operatorname{erfc}\left(\sqrt{\left(\frac{V'}{4} - \beta'\right)\tau} \right) \right] \\ + \frac{(V' - 2)}{2(V'1)} e^{-(V' - 1)\tau} \left[\operatorname{erfc}\left(\left(\frac{V' - 2}{2} \right) \sqrt{\tau} \right) \right] \\ + \frac{(V' - \beta' - 1)V'}{2\beta'(V' - 1)} \operatorname{erfc}\left(\frac{V'}{2}\sqrt{\tau}\right) \\ \left(\frac{-1}{\sqrt{\frac{V'^2}{4} - \beta' + \frac{V'}{2}}} + 1 \right) e^{-\beta'\tau} + \frac{(V' - 2)}{(V' - 1)} e^{-(V' - 1)\tau} \end{array} \right\}$$
$$- \frac{\alpha\rho L}{2k} \left[\left(\operatorname{erfc}\left(-\frac{V'\tau}{2\sqrt{\tau}} \right) - (1 + V'^2\tau) \operatorname{erfc}\left(\frac{V'\tau}{2\sqrt{\tau}} \right) \right) + \frac{2V'\tau}{\sqrt{\pi}} e^{-\left(\frac{V'^2}{4\tau} + \frac{V'^2\tau}{4} \right)} \right] \tag{5.53}$$

The dimensionless temperature can be defined as:

$$T^* = k\delta \frac{T(0,\tau)}{I_1} \tag{5.54}$$

Equation 5.54 is used to obtain the dimensionless surface temperature distribution for exponentially varying time pulses. Equation 5.54 can be used to form the basis for a more accurate solution, which can be obtained by an iterative procedure. It is expected that this solution would be obtained by developing the solution from time zero. In the initial stages, the evaporation rates will be small and thus the solution will be that for the conduction limited heating process (pure conduction process). As the surface temperature rises, so will the evaporation rates. The values for the velocity and the surface temperature can be obtained by stepping forward in time using time steps that are small enough so that the change in surface velocity between steps is small and therefore the velocity derived in the previous step can be used directly in Eq. 5.53. With this new value of the surface temperature, an improved estimate of the surface recession velocity (V) can be obtained and the iteration repeated to give a converged solution.

5.3 Numerical Treatment of Evaporation

Laser high intensity beam interaction with the solid surface results in rapid evaporation of the surface. Depending on the power intensity and the duration of the laser pulse, the pressure generated at the laser irradiated surface becomes very high. This situation is especially true for nanosecond laser pulses. Moreover, the prediction of recoil pressure is essential, since the mass removal rate depends on the pressure differential at the cavity surface. Consequently, one of the governing physical parameters involving laser ablation is the pressure generated at the interface of vapor–liquid phases in the cavity. Since the process is involved with multi-physics problem, analytical solution to the surface evaporation limits the accuracy of the findings due to the several assumptions made for the simplicity. However, it is possible to eliminate some these assumptions in the numerical study and the predictions become more realistic than the analytical results in terms of the comparison with the experimental data. In addition, prediction of pressure in the laser produced cavity is possible through the method of numerical simulations.

The analysis related to the numerical treatment of the surface evaporation due to the laser heating pulse is presented in line with the previous study [2]. The energy equation for each phase needs to be solved independently as well as coupled across the interfaces of the two-phases mutually exists in the mushy zones. In the initial stage of heating, conduction in solid with convective boundary at the surface is considered. The laser output power intensity distribution at the surface of the workpiece is Gaussian and this is incorporated in the analysis such that its center is at the center of the co-ordinate system (Fig. 5.1). The temporal variation of laser power intensity resembling the actual laser pulse is used in the analysis. This arrangement results in an axisymmetric heating of the substrate material.

The diffusion equation for a solid phase heating due to a laser irradiation pulse with a Gaussian intensity profile can be written as:

$$\rho_s Cp_s \frac{\partial T}{\partial t} = \frac{k_s}{r}\frac{\partial}{\partial r}\left(r\frac{\partial T}{\partial r}\right) + k_s\frac{\partial^2 T}{\partial z^2} + S_o \tag{5.55}$$

where ρ_s is the density of the solid phase, Cp_s is the specific heat capacity of the solid phase, k_s is the thermal conductivity of the solid phase, and S_o is the volumetric source term and it is,

$$S_o = I_o\delta(1 - r_f)\exp(-\delta z)\exp\left(-\frac{r^2}{a^2}\right) \tag{5.56}$$

I_o, δ, r_f and a are the laser peak power intensity, absorption coefficient, reflectivity and the Gaussian parameter, respectively.

Initially, the substrate material is assumed at a uniform temperature T_0, i.e.:

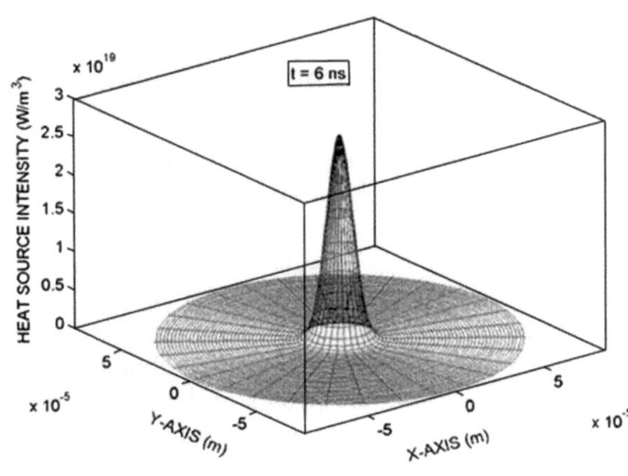

Fig. 5.1 Spatial distribution of laser pulse intensity at the surface of the substrate material after 6 ns

At time zero $\Rightarrow t = 0 : T(r, z, 0) = T_0$ (specified).

At a distance considerably away from the surface (at infinity) in the radial direction a constant temperature T_o is assumed. Since the heating has no effect on the temperature rise at a depth of infinity below the surface, temperature is assumed to be constant and equals to the initial temperature of the substrate material in this region. The respective boundary conditions are:

$$r \text{ at infinity } \Rightarrow r = \infty : T(\infty, z, t) = T_o \text{ (specified)}$$

$$z \text{ at infinity } \Rightarrow z = \infty : T(r, \infty, t) = T_o \text{ (specified)}$$

At the symmetry axis, maximum temperature is assumed and the convective boundary condition is considered at the irradiated surface, i.e.:

$$\text{At symmetry axis } \Rightarrow r = 0 : \frac{\partial T(0, z, t)}{\partial r} = 0$$

$$\text{At the surface } \Rightarrow z = 0 : \frac{\partial T(r, 0, t)}{\partial z} = h(T(r, 0, t) - T_o)$$

where h is taken as 10 W/m^2K due to natural convection from the surface [5].

Since the pressure dependent evaporation temperature of the substrate material is not known, it is assumed that the substrate material has single melting and boiling temperatures. Once the phase change initiates, a mushy zone (mutually existence of two-phases) is introduced across two phases. It is assumed that temperature of the substrate material remains the same during the phase change process, but its enthalpy of the irradiated material changes in this region. This situation can be formulated via energy balance in the mushy zone. Consider a differential element in the mushy zone and let x_m is the mass fraction of liquid present in the element. Energy content (ΔU) of the differential element with volume $\Delta\forall$ at melting temperature T_m can be written as:

$$\Delta U = \rho_m \Delta\forall \left[x_m \left(L_m + c_{p_m}\left(T_m - T_{ref}\right)\right) + c_{p_s}(1 - x_m)\left(T_m - T_{ref}\right)\right] \quad (5.57)$$

where ρ_m is the density of the melt, Cp_m is the specific heat capacity of the melt, T_m is the melting temperature, T_{ref} is the reference temperature, and x_m is:

$$x_m = \frac{m_m}{m_m + m_s} \quad (5.58)$$

T_{ref}, x_m, m_m, m_s are reference temperature for enthalpy, quality of liquid, mass of liquid and mass of solid in the element, respectively. After assuming specific heat of melt is the same as the solid at the melting temperature $\left(c_{p_s} = c_{p_m} \text{ at } T = T_m\right)$, and differentiation of Eq. 5.57 with time yields:

$$\frac{\partial u}{\partial t} = \rho_m L_m \frac{\partial x_m}{\partial t} \quad (5.59)$$

since $c_{p_m}\left(T_m - T_{ref}\right) = \text{const.}$ and $u = \frac{U}{\forall}$.

Substituting Eq. 5.59 into 5.55 and re-arrangement results the energy equation for the differential element in the mushy zone:

$$\rho_m L_m \frac{\partial x_m}{\partial t} = \frac{k_m}{r}\frac{\partial}{\partial r}\left(r\frac{\partial T}{\partial r}\right) + k_m\frac{\partial^2 T}{\partial z^2} + S_o \tag{5.60}$$

Equation 5.60 is valid in the mushy zone where $0 \leq x_m \leq 1$, i.e. temperature of the cells with $0 \leq x_m \leq 1$ is set to melting temperature$(T = T_m)$. For $x_m > 1$, Eq. 5.55 is used to determine the temperature rise in the liquid heating with using the liquid thermal properties in the equation. Moreover, the liquid heating continues until the boiling point is reached in the substrate material; in which case, a new mushy zone is formed. In this case, Eq. 5.60 is modified for a differential element in the mushy zone subjected to evaporation, i.e.:

$$\rho_b L_b \frac{\partial x_b}{\partial t} = \frac{k_b}{r}\frac{\partial}{\partial r}\left(r\frac{\partial T}{\partial r}\right) + k_b\frac{\partial^2 T}{\partial z^2} + S_o \tag{5.61}$$

where ρ_b is the density of the vapor phase, k_b is the thermal conductivity of the vapor phase, and x_b is the fraction of vapor phase.

Equation 5.61 is applicable for temperature $T = T_b$ and $0 \leq x_b \leq 1$ in the mushy zone (partially liquid partially vapor zone); in which case, temperature of the cells with $0 \leq x_b \leq 1$ is set to the boiling temperature of the substrate material$(T = T_b)$. It should be noted that x_m is replaced with x_b in Eq. 5.60, which represents the fraction of vapor phase in the differential element. The calculation of x_b is the same as x_m, provided that latent heat of fusion is replaced with latent heat of evaporation of the substrate material in Eq. 5.61 in the later.

The boundary condition at the evaporating surface is introduced in relation to Eq. 5.61. In this case, the temperature along the evaporated surface is kept at boiling temperature of the substrate material, i.e., the cells in the evaporated region are kept at boiling temperature, i.e.:

In the mushy zone, at $z = z_b \Rightarrow T(r, z_b, t) = T_b$

where z_b represents the axial location at the evaporated surface.

Equations 5.60 and 5.61 provide the relative position of solid–liquid and liquid–vapor interface in the substrate material. Liquid–vapor interface determines the shape and size of the cavity generated during evaporation process.

The vapor jet emanating from the laser produced cavity is modeled numerically using a control volume approach and the fluid dynamic/mass transfer model is incorporated in the analysis. The time-varying cavity shape, mass flux of the vapor and the temperature distribution at the cavity surface are considered to be the inputs for the simulations. Since the flow situation is very complicated due to the involvement of transiently developing jet and recessing cavity surface, the absorption of laser beam by the vapor front is omitted in the flow analysis for the simplicity. In the flow analysis, the time averaged conservation equations are incorporated for an unsteady, incompressible, axisymmetric turbulent flow situation resembling the vapor jet expansion. The Standard k-ε turbulence model is used to account for the turbulence. The species transport model is also used to account for the mass transfer of the vapor jet from the cavity into the stagnant ambient. It should be noted that all the unknown quantities are the time-averaged due to the turbulence modeling.

Continuity Equation:

$$\frac{1}{r}\frac{\partial(rV_r)}{\partial r} + \frac{\partial V_z}{\partial r} = 0 \tag{5.62}$$

Radial momentum:

$$\frac{\partial(\rho V_r)}{\partial t} + \frac{1}{r}\frac{\partial\left(\rho r V_r^2\right)}{\partial r} + \frac{\partial(\rho V_r V_z)}{\partial z} = -\frac{\partial p}{\partial r} + \frac{2}{r}\frac{\partial}{\partial r}\left(\mu_{eff} r \frac{\partial V_r}{\partial r}\right) + \frac{\partial}{\partial z}\left(\mu_{eff}\frac{\partial V_r}{\partial z}\right) + \frac{\partial}{\partial z}\left(\mu_{eff}\frac{\partial V_z}{\partial z}\right) - 2\mu_{eff}\frac{V_r}{r^2} \tag{5.63}$$

Axial momentum:

$$\frac{\partial(\rho V_z)}{\partial t} + \frac{1}{r}\frac{\partial(\rho r V_r V_z)}{\partial r} + \frac{\partial\left(\rho V_z^2\right)}{\partial z} = -\frac{\partial p}{\partial z} + \frac{1}{r}\frac{\partial}{\partial r}\left(\mu_{eff} r \frac{\partial V_z}{\partial r}\right) + 2\frac{\partial}{\partial z}\left(\mu_{eff}\frac{\partial V_z}{\partial z}\right) + \frac{1}{r}\frac{\partial}{\partial r}\left(\mu_{eff} r \frac{\partial V_r}{\partial z}\right) \tag{5.64}$$

where,

$$\mu_{eff} = \mu + \mu_t \ : \ \mu_t = \frac{\rho C_\mu K^2}{\varepsilon} \ : \ C_\mu = 0.09$$

here μ_{eff} is the effective viscosity, μ is the bulk viscosity, μ_t is the turbulent viscosity.

Energy Equation:

$$\begin{aligned}\frac{\partial(\rho E)}{\partial t} + \frac{1}{r}\frac{\partial(rV_r\rho E)}{\partial r} + \frac{\partial(V_z\rho E)}{\partial z} &= \frac{1}{r}\frac{\partial}{\partial r}\left(rk_{eff}\frac{\partial T}{\partial r}\right) + \frac{\partial}{\partial z}\left(k_{eff}\frac{\partial T}{\partial z}\right)\\ &+ \left[\frac{1}{r}\frac{\partial}{\partial r}\left(rh_{vapour}\left(\rho D + \frac{\mu_t}{Sc_t}\right)\frac{\partial Y_{vapour}}{\partial r}\right) + \frac{\partial}{\partial z}\left(h_{vapour}\left(\rho D + \frac{\mu_t}{Sc_t}\right)\frac{\partial Y_{vapour}}{\partial z}\right)\right]\\ &+ \left[\frac{1}{r}\frac{\partial}{\partial r}\left(rh_{air}\left(\rho D + \frac{\mu_t}{Sc_t}\right)\frac{\partial Y_{air}}{\partial r}\right) + \frac{\partial}{\partial z}\left(h_{air}\left(\rho D + \frac{\mu_t}{Sc_t}\right)\frac{\partial Y_{air}}{\partial z}\right)\right]\end{aligned} \tag{5.65}$$

where, $E = Y_{air}h_{air} + Y_{vapor}h_{vapor}$, after neglecting the contribution of kinetic energy. Enthalpy of vapor and water are:

$$h_{vapor} = \int_{T_{ref}}^{T} Cp_{vapor} dT = Cp_{vapor}\left(T - T_{ref}\right) \tag{5.66}$$

and

$$h_{water} = \int_{T_{ref}}^{T} Cp_{water} dT = Cp_{water}\left(T - T_{ref}\right) \tag{5.67}$$

In Eqs. 5.65–5.67, the properties and parameters are:

$k_{eff} = k + k_t \ : \ k_t = Cp\frac{\mu_t}{\mathrm{Pr}_t}$ and

$D = 2.88 \times 10^{-5} \left(m^2/s\right) \ : Sc_t = 0.7 : \mathrm{Pr}_t = 0.85 \ : \ T_{ref} = 298.15\,K$

where k_{eff} is the effective thermal conductivity, k is the bulk thermal conductivity, k_t is the turbulent contribution to thermal conductivity, D is the diffusion coefficient, Sc_t is the Schmidt number.

Turbulence Kinetic Energy Equation, K:

$$\frac{\partial(\rho K)}{\partial t} + \frac{1}{r}\frac{\partial(\rho r V_r K)}{\partial r} + \frac{\partial(\rho V_z K)}{\partial z} = \frac{1}{r}\frac{\partial}{\partial r}\left(\frac{\mu_{eff}}{\sigma_K} r \frac{\partial K}{\partial r}\right) + \frac{\partial}{\partial z}\left(\frac{\mu_{eff}}{\sigma_K}\frac{\partial K}{\partial z}\right) - \rho\varepsilon + P_K \tag{5.68}$$

where,

$$P_K = \mu_{eff}\left[2\left\{\left(\frac{\partial V_z}{\partial r}\right)^2 + \left(\frac{\partial V_r}{\partial r}\right)^2 + \left(\frac{V_r}{r}\right)^2\right\} + \left(\frac{\partial V_z}{\partial r} + \frac{\partial V_r}{\partial z}\right)^2\right] \tag{5.69}$$

Rate of Dissipation Equation, ε:

$$\frac{\partial(\rho\varepsilon)}{\partial t} + \frac{1}{r}\frac{\partial(\rho r V_r \varepsilon)}{\partial r} + \frac{\partial(\rho V_z \varepsilon)}{\partial z} = \frac{1}{r}\frac{\partial}{\partial r}\left(\frac{\mu_{eff}}{\sigma_\varepsilon} r \frac{\partial \varepsilon}{\partial r}\right) + \frac{\partial}{\partial z}\left(\frac{\mu_{eff}}{\sigma_\varepsilon}\frac{\partial \varepsilon}{\partial z}\right) - C_1\frac{\varepsilon}{K}P_K - C_2\rho\frac{\varepsilon^2}{K} \tag{5.70}$$

where,

$$P_K = \mu_{eff}\left[2\left\{\left(\frac{\partial V_z}{\partial r}\right)^2 + \left(\frac{\partial V_r}{\partial r}\right)^2 + \left(\frac{V_r}{r}\right)^2\right\} + \left(\frac{\partial V_z}{\partial r} + \frac{\partial V_r}{\partial z}\right)^2\right] \tag{5.71}$$

and

$$\sigma_K = 1,\ \sigma_\varepsilon = 1.3,\ C_1 = 1.44,\ C_2 = 1.92$$

Species Transport Equation:

$$\frac{\partial(\rho Y_{vapour})}{\partial t} + \frac{1}{r}\frac{\partial(rV_r\rho Y_{vapour})}{\partial r} + \frac{\partial(V_z\rho Y_{vapour})}{\partial z} = \begin{bmatrix} \frac{1}{r}\frac{\partial}{\partial r}\left(r\left(\rho D + \frac{\mu_t}{Sc_t}\right)\frac{\partial Y_{vapour}}{\partial r}\right) \\ +\frac{\partial}{\partial z}\left(\left(\rho D + \frac{\mu_t}{Sc_t}\right)\frac{\partial Y_{vapour}}{\partial z}\right) \end{bmatrix} \tag{5.72}$$

where $Y_{air} = 1 - Y_{vapour}$.

Initial and Boundary Conditions:

Symmetry Axis: $(r = 0)$

At the symmetry axis all the unknown quantities are considered to be the maximum accept the r-direction velocity, which is zero.

$$\left.\frac{\partial V_z}{\partial r}\right|_{\substack{z=z\\r=0}} = 0 : V_r(z,0) = 0 : \left.\frac{\partial T}{\partial r}\right|_{\substack{z=z\\r=0}} = 0 : \left.\frac{\partial K}{\partial r}\right|_{\substack{z=z\\r=0}} = 0$$

$$: \left.\frac{\partial \varepsilon}{\partial r}\right|_{\substack{z=z\\r=0}} = 0 : \left.\frac{\partial Y_{vapor}}{\partial r}\right|_{\substack{z=z\\r=0}} = 0$$

Outflow: (at $z = 0$)

At the outflow boundary perpendicular to the z-axis the normal derivatives of all the unknown quantities are considered to be zero accept the r-direction velocity, whose value is zero as required from the continuity equation.

$$\left.\frac{\partial V_z}{\partial z}\right|_{\substack{z=0\\r=r}} = 0 : V_r(0,r) = 0 : \left.\frac{\partial T}{\partial z}\right|_{\substack{z=0\\r=r}} = 0 : \left.\frac{\partial K}{\partial r}\right|_{\substack{z=0\\r=r}} = 0$$

$$: \left.\frac{\partial \varepsilon}{\partial r}\right|_{\substack{z=0\\r=r}} = 0 : \left.\frac{\partial Y_{vapor}}{\partial r}\right|_{\substack{z=0\\r=r}} = 0$$

Outflow: (at $r = r_{max}$)

At the outflow boundary perpendicular to the r-axis the normal derivatives of all the unknown quantities are considered to be zero accept the z-direction velocity, whose value is zero as required from the continuity equation.

$$V_z(z, r_{\max}) = 0 : \left.\frac{\partial V_r}{\partial r}\right|_{\substack{z=z\\r=r_{\max}}} = 0 : \left.\frac{\partial T}{\partial r}\right|_{\substack{z=z\\r=r_{\max}}} = 0 : \left.\frac{\partial K}{\partial r}\right|_{\substack{z=z\\r=r_{\max}}} = 0$$

$$: \left.\frac{\partial \varepsilon}{\partial r}\right|_{\substack{z=z\\r=r_{\max}}} = 0 : \left.\frac{\partial Y_{vapor}}{\partial r}\right|_{\substack{z=0\\r=r_{\max}}} = 0$$

Solid Wall: $z = f(r,t)$

The surface of the substrate material including the cavity wall acts like a wall in the solution domain and hence a no-slip and no-temperature jump boundary conditions are considered. At the cavity surface, the vapor mass fraction is considered to be one whereas the water mass fraction is considered to be zero. In this case, the function $f(r, t)$ defining the cavity wall shape, as obtained from the heat transfer analysis, can be presented in algebraic form, i.e.:

$$f(r, t) = \left[c\left((r/\Delta r)^2 - r_{\max}^2\right) + d\left((r/\Delta r)^3 - r_{\max}^3\right)\right](\rho_l/\rho_v)\Delta z \tag{5.73}$$

where

$$c = cc_3 + cc_2(t/\Delta t) + cc_1(t/\Delta t)^2$$

$$cc_1 = 4.4499 \times 10^{-6} \; : \; cc_2 = -0.0026726 \; : \; cc_3 = -0.15077$$

and

$$d = cd_3 + cd_2(t/\Delta t) + cd_1(t/\Delta t)^2$$

$$cd_1 = -1.7974 \times 10^{-7} \; : \; cd_2 = 0.00011187 \; : \; cd_3 = 0.0062477$$

and

$$r_{\max} = cr_5 + cr_4(t/\Delta t) + cr_3(t/\Delta t)^2 + cr_2(t/\Delta t)^3 + cr_1(t/\Delta t)^4$$

where

$$cr_1 = -2.5982 \times 10^{-9} \; : \; cr_2 = 2.0455 \times 10^{-6} \; : \; cr_3 = -0.00059955$$
$$: \; cr_4 = 0.086871 \; : \; cr_5 = 7.7763$$

The numerical values for the space increments are:

$$\Delta z = 3.2415 \times 10^{-8} \text{ m}$$
$$\Delta r = 8.3333 \times 10^{-8} \text{ m}$$

The recession velocity of the cavity wall as obtained from the above analysis can be represented in the algebraic form. In this case, the cavity size is limited with $0 \le r \le r_{\max}$. Therefore, the recession velocity of the cavity along the vertical direction is:

$$V_z(f(r, t), r) = \frac{\rho_l}{\rho_v}\frac{\partial f}{\partial t}\frac{\partial f/\partial r}{\sqrt{1 + (\partial f/\partial r)^2}} \quad \text{for } 0 \le r \le r_{\max} \tag{5.74}$$

Moreover, outside of the cavity a stationary solid wall is considered. The recession velocity along the vertical axis is, therefore:

$$V_z(f(r,t), r) = 0 \quad \text{for } r > r_{\max} \tag{5.75}$$

The recession velocity of the cavity along the radial direction is:

$$V_r(f(r,t),r) = -\frac{\rho_l}{\rho_v}\frac{\partial f}{\partial t}\frac{1}{\sqrt{1+(\partial f/\partial r)^2}} \quad \text{for } 0 \le r \le r_{\max} \tag{5.76}$$

Outside of the cavity a stationary solid wall is considered. The recession velocity along the radial direction is, therefore:

$$V_r(f(r,t),r) = 0 \quad \text{for } r > r_{\max} \tag{5.77}$$

Temperature at the cavity wall is determined from the heat transfer analysis and can be presented in algebraic form. In this case, temperature at the cavity wall is the same as the boiling temperature of the substrate material (T_b), i.e.:

$$T(f(r,t),r) = T_b \quad \text{for } 0 \le r \le r_{\max}$$

$$T(f(r,t),r) = 2834\, exp\left(-btemp((r/\Delta r) - r_{\max})^2\right) + 300 \quad \text{for } r > r_{\max}$$

where

$$btemp = ctemp(t/\Delta t)^2 + dtemp(t/\Delta t) + etemp \; : \; ctemp = 1.0370442955011 \times 10^{-6}$$

$$dtemp = -3.39682806506743 \times 10^{-5} \; : \; etemp = 0.0241815719639816$$

Turbulence kinetic energy and dissipation of the jet in the cavity wall region is assumed to be constant and taken as:

$$K(f(r,t),r) = 1\,(\text{m/s})^2 \text{ and } \varepsilon(f(r,t),r) = 1\,(\text{m/s})^2$$

$$Y_{vapour}(f(r,t),r) = 1 \quad \text{for } 0 \le r \le r_{\max}$$

$$Y_{vapour}(f(r,t),r) = 0 \text{ for } r > r_{\max}$$

Initial Conditions:

Initially the ambient water is assumed as stagnant; therefore, the z and r-directions velocity components are zero. Initially, temperature is considered to be uniform and equal to 300 K in water ambient and the vapor mass fraction is zero whereas the water mass fraction is one throughout the domain.

$$V_z(z,r) = 0 \; : \; V_r(z,r) = 0 \; : \; T(z,r) = 300\,K \; : \; K(z,r) = 1 \; : \; \varepsilon(z,r) = 1 \; : \; Y_{vapour}(z,r) = 0$$

5.4 Numerical Solution of Governing Equations

The numerical solution of the governing equations is divided into two parts. The first part deals with the solution of the phase change problem during the laser irradiation pulse and the second part provides the solution for the vapor jet emanating from the laser produced cavity.

5.4.1 Phase Change Process

Equation 5.55 is applicable to solid and liquid heating, Eq. 5.60 is applicable to a mushy zone at solid–liquid interface and Eq. 5.61 is applicable to mushy zone at liquid–vapor interface. To discretize the governing equations, a finite difference scheme is introduced. The details of the numerical scheme are given in [6]. To compute the equations discretized for temperature field and relative positions of solid–liquid and liquid–vapor interface, an implicit scheme is used, i.e. using the initial conditions, the temperature in the whole domain is calculated for following time steps with the respective conditions.

The calculation domain is divided into grids and grid independence test is performed for different grid size and orientation. The grid size resulting grid independent solution is used, which is 100×120 mesh points in the r and z-axes. A computer program based on implicit scheme is developed to compute the temperature field. It should be noted that the laser pulse properties employed in the simulations resemble the actual pulse used in the experiment. The model study is applicable for pulse laser with spatial power intensity distribution is Gaussian; in this case, the model predicts the results, which are suitable for the pulsed laser processing of metallic substrates such as Nd:YAG laser applications.

5.4.2 Transiently Developing Vapor Jet

A control volume approach is incorporated when discretizing the governing equations [7]. A staggered grid arrangement is used in which the velocities are stored at a location midway between the grid points, i.e. on the control volume faces. All other variables including pressure are calculated at the grid points. This arrangement gives a convenient way of handling the pressure linkages through the continuity equation and is known as Semi-Implicit Method for Pressure-Linked Equations (SIMPLE) algorithm. The details of this algorithm are given in [7].

The computer program used for the present simulation can handle a non-uniform grid spacing. Along the radial direction fine uniform grid spacing is allocated at the inlet (in cavity symmetry axis region) while gradually increasing spacing is considered away from the inlet (in the cavity edge region). Along the axial direction, again fine uniform grid spacing is used inside and near the cavity while the grid spacing gradually increases away from the cavity. The number of grid points in the radial direction is 300 while 215 grid points are used in the axial direction. Since the problem is involved with the moving boundary, the moving meshes are accommodated in the cavity to account for the cavity recession with time. The grid independence test is conducted and grid size (215 × 300) resulting in grid independent solution is used in the simulations

Eight variables are computed at all grid points. These are the two velocity components, local pressure, two turbulence quantities, temperature and two mass fractions.

5.5 Discussions

The discussions are divided into two sub-sections, which include the findings of the analytical solution to surface evaporation and the predictions of the numerical treatment of phase change at the surface and vapor jet flow from the cavity. The discussions are given under the appropriate sub-headings:

5.5.1 Solution of Analytical Treatment of Evaporation

A closed form solution for surface temperature distribution due to time exponentially decaying laser pulse intensity is obtained using a Laplace transformation method. The findings are discussed in line with the previous study [1]. The pulse parameters and material properties used in the analysis are given in Table 5.1.

To compare the closed form solution with the analytical solution obtained from the previous study for the conduction limited heating case, the present closed form solution should reduce to the solution for the conduction limited heating case. To achieve this, the boundary condition at the surface $\frac{\partial T}{\partial x}\Big|_{x=0} = \frac{\rho V L_{ev}}{k}$ is replaced by $\frac{\partial T}{\partial x}\Big|_{x=0} = 0$ and V' is set to zero in Eq. 5.53. The details of the mathematical arrangements are given in [1]. Hence, the non-dimensional surface temperature for the conduction limited heating case yields:

$$\frac{T(0,\tau)}{\frac{I_1}{k\delta}} = \frac{1}{(\beta' + 1)}\left\{\frac{2}{\sqrt{\pi}\sqrt{\beta'}}F\left[\sqrt{\beta'\tau}\right] + e^{\tau}\left[\operatorname{erfc}(\sqrt{\tau})\right] - e^{-\beta'\tau}\right\} \tag{5.78}$$

where:

$$F(\chi) = e^{-\chi^2}\int_0^{\chi} e^{v^2}\,dv \tag{5.79}$$

Table 5.1 Properties of the substrate material used in the analytical solutions

$\delta \times 10^7$ (1/m)	$\alpha \times 10^{-4}$ (m²/s)	c_p (J/kgK)	ρ (kg/m³)	k (W/mK)	β (1/s)	γ (1/s)	I_1 (W/m²)
6.16	0.224	460	7,880	80.3	10^7–10^9	0.3	10^{13}

$F(\chi)$ is Dawson's integral. Eq. 5.78 is identical to the closed form solution obtained in the previous study [3].

Figure 5.2 shows temporal variation of surface temperature for different values of pulse parameter (β') as well as predictions of the numerical simulation. It can be observed that the predictions of numerical simulation agree well with the closed form solution. Temperature rises rapidly in the early heating period and the rate of temperature rise reduces as the heating period progresses. In this case, the internal energy gain of the substrate material is high in the surface vicinity due to absorption of laser energy. Since the temperature in this period is less than the melting temperature of the substrate material ($T^* < 0.85$), conduction limited heating takes place. The internal energy gain dominates over the conduction losses; in which case, temperature in the surface vicinity increases rapidly and the increase in temperature gradient is not considerably large to transfer high rate of energy via diffusion from surface vicinity to the solid bulk. This, in turn, suppresses temperature rise in this region. As the heating period increases, the rate of temperature rise reduces. This occurs because of one or all of the followings: (1) the temperature gradient becomes high, which in turn accelerates the energy diffusion from surface vicinity to solid bulk, therefore, the temperature rise in the surface vicinity is suppressed, and (2) power intensity decreases exponentially with time, which results in less power being absorbed by the substrate material, and (3) some portion of the energy from the substrate surface is transferred by the evaporation process, which reduces temperature rise in this region. The influence of pulse parameter (β') on the temperature rise is significant, which is more pronounced for the heating period $\tau > 0.15$. In this case, increasing β' lowers the temperature rise due to low power intensity available at the surface.

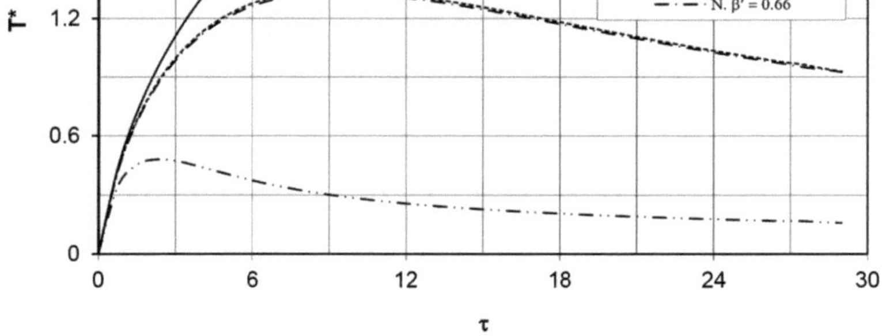

Fig. 5.2 Temporal variation of dimensionless surface temperature obtained from closed form solution and numerical simulation (presented as $N \cdot \beta'$)

Figure 5.3 shows the maximum temperature and corresponding pulse parameter β' The maximum temperature reduces with increasing β', except at low values of β'at which it reduces slightly. In this case, the rate of evaporation suppresses the rise of the maximum temperature at the surface, since low β' results in high power intensity in the early heating period. Beside the low value of β', the maximum temperature decays almost linearly with increasing β'. Therefore, β' can be selected to obtain the maximum desired temperature at the surface.

Figure 5.4 shows the time of maximum temperature with pulse parameter β'. The time corresponding to the maximum temperature reduces with increasing β'. This reduction is sharp at low values of β' and it becomes gradually with increasing β'. This indicates that the time corresponding to the maximum temperature changes sharply once the rate of evaporation increases. Therefore, increasing power intensity in the early heating period changes the time occurrence of maximum temperature drastically. at low β' values.

5.5.2 Predictions from Numerical Treatment of Evaporation

Laser heating of steel surface is considered and phase change process in the irradiated region is modeled. The flow field in the laser produced cavity is obtained through numerical solution of the governing flow equations. k-ε turbulence model is incorporated to account for the turbulence. The discussions made on the numerical treatment of evaporation at the surface due to laser pulse is presented in line with the previous study [2]. The material properties and pulse intensity used in

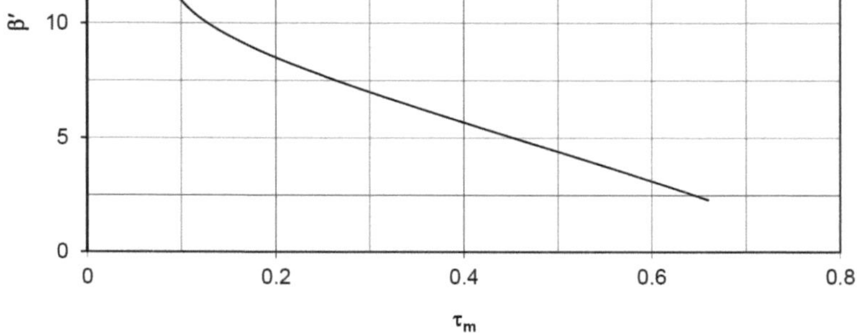

Fig. 5.3 Pulse parameter (β') with time corresponding to maximum surface temperature

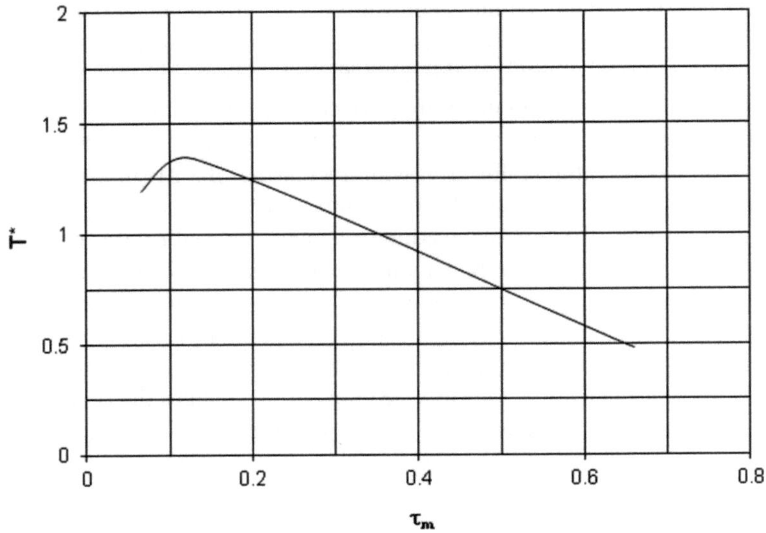

Fig. 5.4 The maximum temperature with time corresponding to maximum surface temperature

Table 5.2 Thermal properties of steel used in the simulations for phase change

T_m (K)	T_b (K)	$\delta \times 10^7$ (1/m)	Cp (J/kgK)	ρ (kg/m^3)	k (W/mK)	L_m (J/kg)	L_{ev} (L_b)(J/kg)
1,810	3,134	6.16	420	7,860	63	247,112	6,213,627

Table 5.3 The properties of the laser power intensity used in the simulations

Peak intensity (W/m^2)	Gaussian parameter (1/m)	Nominal pulse length (ns)
9×10^{12}	120,000	24

Table 5.4 Properties of ambient (water) and vapor front used in the simulations

	Density	Viscosity	Mass diffusivity	Specific heat	Thermal conductivity	Molecular weight
Substance	ρ(1/m)	v(kg/m/s)	D(m^2/s)	Cp(J/kgK)	k(W/mK)	M(kg/kmol)
Ambient (Water)	998.2	0.001003	2.88×10^{-5}	4,182	0.6	18.0152
Vapor front	50,800	1.34×10^{-5}		2,014	0.0261	56

the simulations of the phase change are given Tables 5.2 and 5.3 respectively. The properties used in the simulations of vapor jet are given in Table 5.4.

Figure 5.5 shows the cavity depth along the radial direction measured from SEM micrographs and predictions. The cavity profiles obtained from the experiment and the predictions are found to be in a good agreement. The differences in

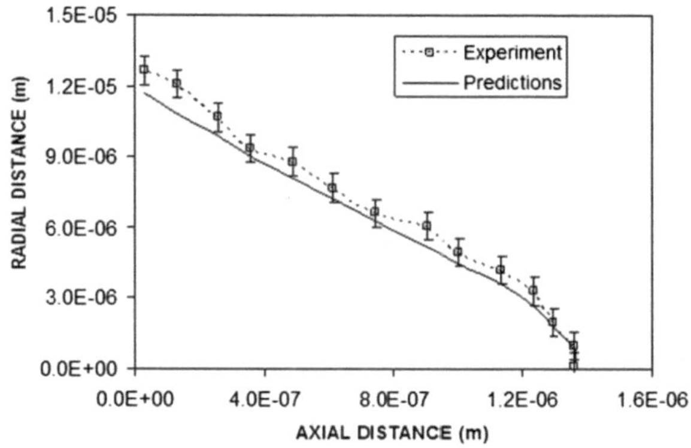

Fig. 5.5 Laser produced cavity profiles obtained from predictions and experiment. The radial distance of zero represents the symmetry axis

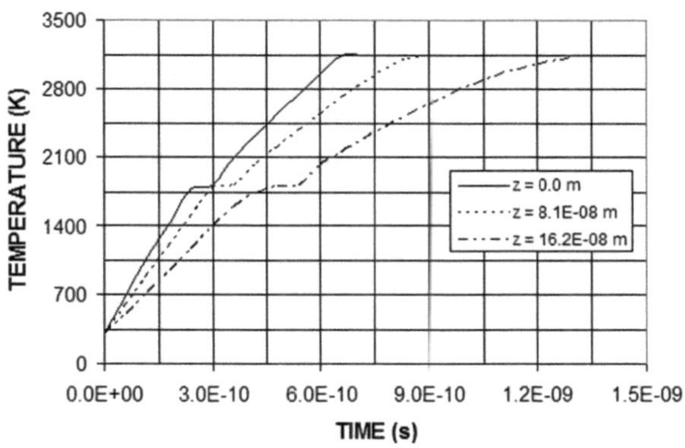

Fig. 5.6 Temporal variation of temperature at different depths below the surface and radial location is the symmetry axis

both profiles are considerably small along the symmetry axis (radial location is zero); however, some small discrepancies occur in the region close to the cavity edge. This is because of the measurement errors and the assumptions made in the analysis. Nevertheless, the discrepancy is small.

Figure 5.6 shows temporal variation of temperature at different depths inside the substrate material where the radial location is the symmetry axis. Surface temperature rises rapidly in the solid phase $(T \leq T_m)$ and phase change due to liquid formation results in a steady temperature with progressing time. This situation is more pronounced at some depth below the surface where the absorbed laser energy is low due to exponent decay of intensity distribution with depth (Lambert's law). As the time a progress, the liquid phase heating is progressed until temperature reaches to the evaporation temperature of the substrate material. The rate of temperature rise in the liquid phase is smaller than the rate of temperature rise in the solid phase. This is mainly because of the specific heat capacity, which is slightly higher for the liquid phase (Table 5.2). Similarly, at some depth below the surface, the rate of temperature rise in the liquid phase is considerably low, which is due to low energy aborted from the irradiated field at this depth. The time corresponding to reach the vapor temperature at the surface is about 2.4 ns from the pulse beginning. However, as the depth below the surface increases, this time increases, i.e. it is about 4.8 ns at depth equal to the absorption depth of the substrate material ($1/\delta = 16.2 \times 10^{-8}$ m).

Figure 5.7 shows three-dimensional view of laser produced cavity and its cross-section (phases developed in the cavity) for the heating duration of 19.7 ns. The thickness of the mushy zone is smaller at liquid–solid interface than that corresponding to vapor–liquid interface. This is because of the latent heat of fusion, which is smaller than the latent heat of evaporation (Table 5.4). Moreover, the influence of distribution of laser power intensity in the radial direction on the cavity formation is evident. In this case, the liquid layer thickness is small at the edges of the cavity while the thickness of the mushy zone is large in this region. Since the absorption of incident beam takes place along the vertical axis (z-axis), heat diffusion in the radial direction does not contribute significantly to the cavity extension in the radial direction.

Figure 5.8 shows velocity magnitude in the vicinity of the cavity wall (cells next to the cavity wall) in the radial direction for different time periods. Velocity magnitude reaches significantly high values in the vicinity of the cavity, which is particularly true at the symmetry axis where ($r = 0$). High rate of increase in the velocity magnitude is due to high rate of momentum exchange in the vicinity of the cavity wall. The recession velocity of the surface is as high as 100 m/s along the symmetry axis (Fig. 5.9). Consequently high rate of evaporation enables the vapor front velocity reaches as high as 1,000 m/s. It should be noted that the density of the evaporating front in the region next to the cavity wall is assumed to be 1/10 of the liquid density. As the radial distance increases towards the cavity edge, velocity magnitude reduces because of the relatively lower rate of evaporation as compared to symmetry axis. It should be noted that he cavity size increases with progressing time so that the velocity magnitude extends in the radial direction with progressing time. Moreover, in the early eating period velocity magnitude is slightly lower than that corresponding to late heating period. This is because of the recession velocity of the cavity wall, which is slightly lower n the early heating periods so that the momentum exchange between the recessing surface and evaporating front is slightly low resulting in slightly small velocity magnitude.

Figure 5.10 shows velocity magnitude in the radial direction and the axial location (z-axis location) is the cavity exit. Velocity profiles along the radial direction differ slightly from those corresponding to cavity wall region, which is particularly true for early heating periods. This is because of the expansion of the vapor front into stagnant water ambient. It should be noted that stagnant water ambient is considered at the workpiece surface prior to laser irradiation to resemble the laser shock processing. Consequently, vapor front expanding into a stagnant water results in modification in velocity magnitude, particularly at radial location away from the symmetry axis in the early heating period. This situation becomes significant in the region of the symmetry axis for late heating periods. This is because of the axial and radial momentum exchange between the vapor jet

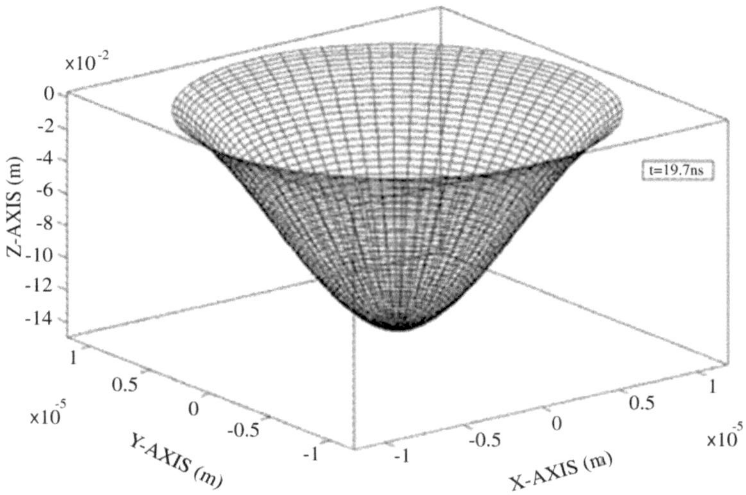

Three-dimensional view of laser produced cavity

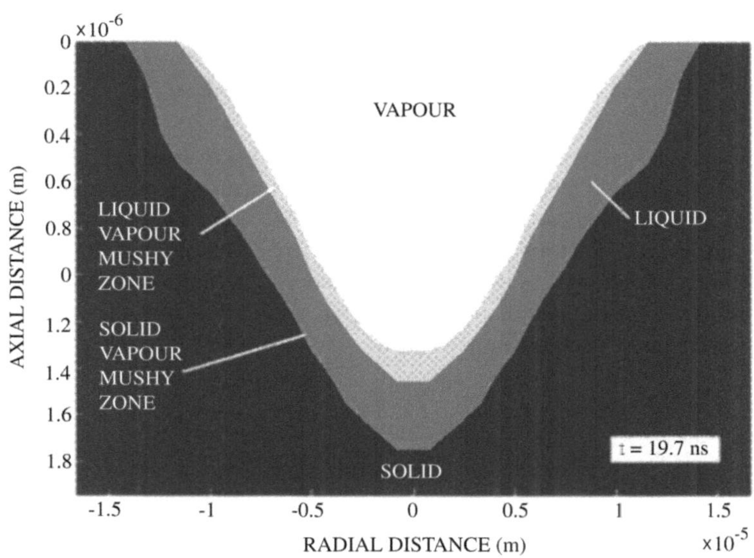

Cross-sectional view of laser produced cavity

Fig. 5.7 Laser produced cavity at 19.7 ns of the heating pulse. Three-dimensional view of laser produced cavity, cross-sectional view of laser produced cavity

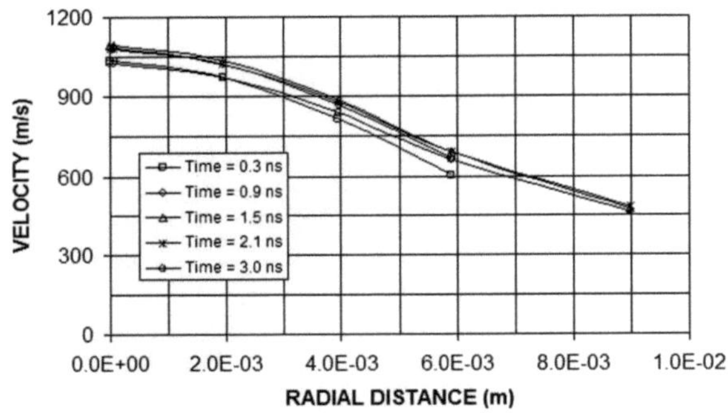

Fig. 5.8 Velocity magnitude along the radial direction in the region close to the cavity wall at different heating durations

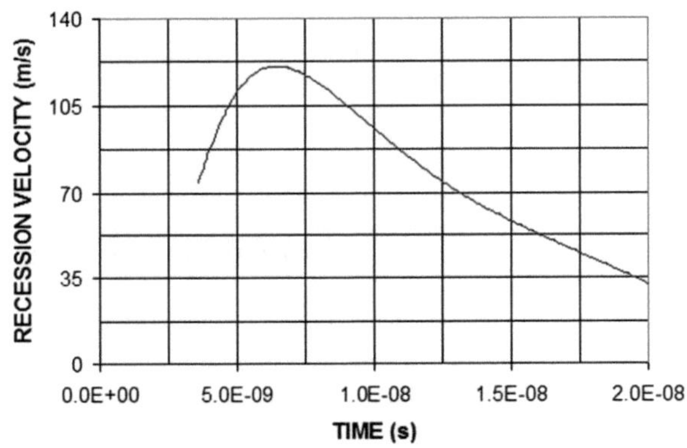

Fig. 5.9 Temporal variation of recession velocity of the cavity along the symmetry axis

and ambient fluid (water). In the early heating period, axial momentum change of the jet is almost a sole mechanism while in the later heating period contribution of radial momentum exchange becomes significant in vapor front expansion. Consequently, immediately after the vapor jet penetration into stagnant water ambient, momentum gain by water is dissipated via its motion in axial and radial directions. This accelerates the vapor jet front penetration into the water, particularly in the region of the symmetry axis.

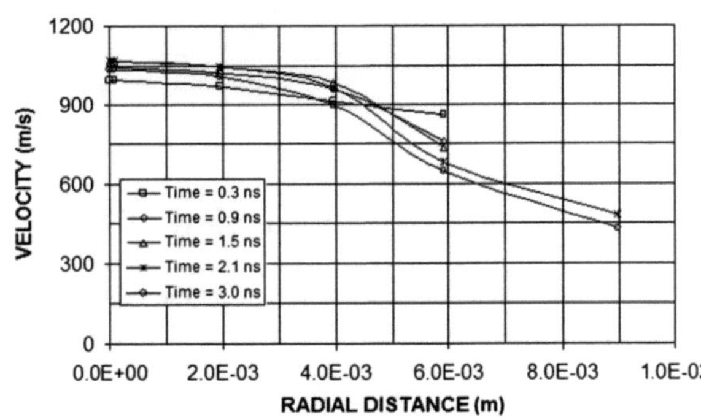

Fig. 5.10 Velocity magnitude along the radial direction in the region of cavity exit at different heating durations

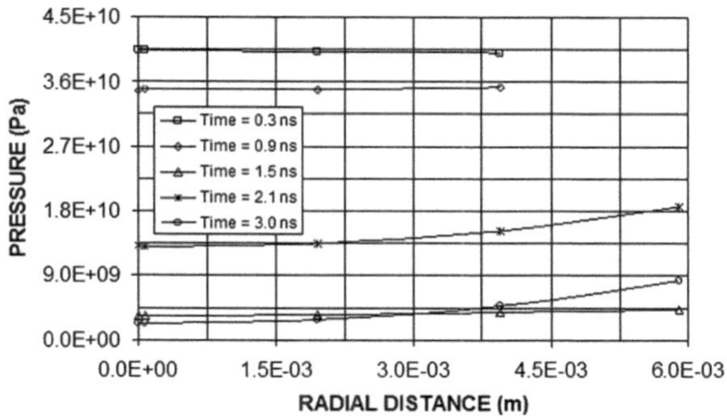

Fig. 5.11 Recoil pressure variation along the radial direction in the region close to the cavity wall at different heating durations

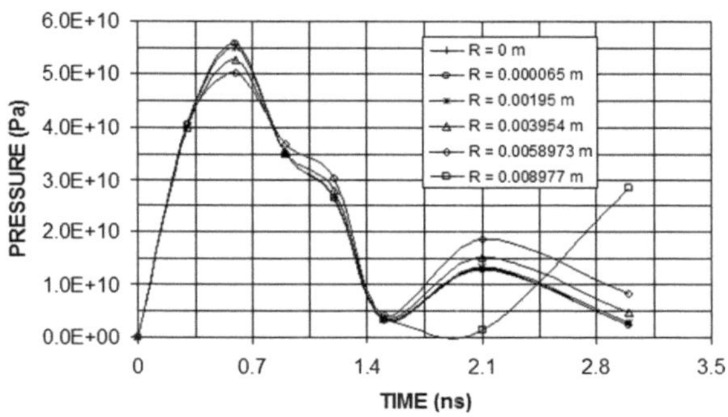

Fig. 5.12 Temporal variation of recoil pressure variation in the region close to the cavity wall at different radial locations

Figure 5.11 shows pressure variation along the radial location in the vicinity of the cavity wall. The pressure in this region is a result of linear momentum change in the axial direction and being called a recoil pressure. Pressure is high in the early heating period and as the time progresses, recoil pressure reduces. This is true for all radial locations. This situation can also be seen from Fig. 5.12 in which temporal variation of recoil pressure is shown. The rapid increase of recoil pressure in the early heating period is associated with the rapid increase in evaporation front velocity in the early heating period. The maximum value of recoil pressure reaches about 5.5×10^{10} Pa. It should be noted that the magnitude of axial velocity is in the order of 1,000 m/s and the density of vapor front is 800 kg/m^3, i.e. the value of pressure reaches above 10^{10} Pa due to linear momentum change. The temporal and radial variation of recoil pressure is associated with the complex flow structure generated in the cavity via expansion of the vapor jet and recession of the cavity wall due to evaporation.

Figure 5.13 shows pressure variation in the radial direction at the cavity exit for different heating durations. Similar to those shown in Fig. 5.12, recoil pressure attains high values during the early heating periods. This situation can also be seen from Fig. 5.14, in which temporal variation of recoil pressure is shown. However, the spatial distribution of pressure becomes slightly different than that corresponding to the cavity wall vicinity. In this case vapor jet front expansion into stagnant water modifies the pressure distribution at the cavity exit. This is true for

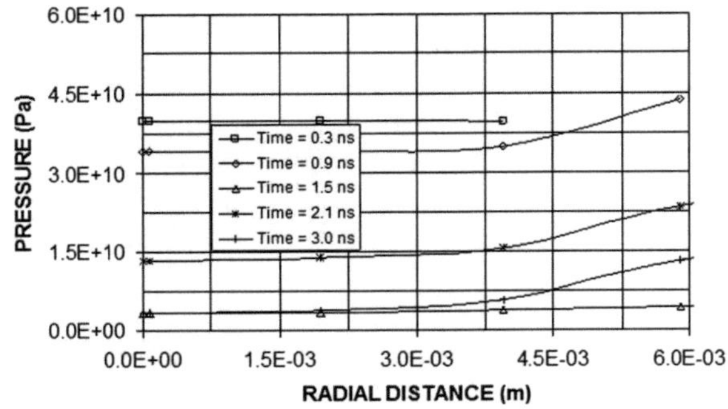

Fig. 5.13 Recoil pressure variation along the radial direction in the region of cavity exit at different heating durations

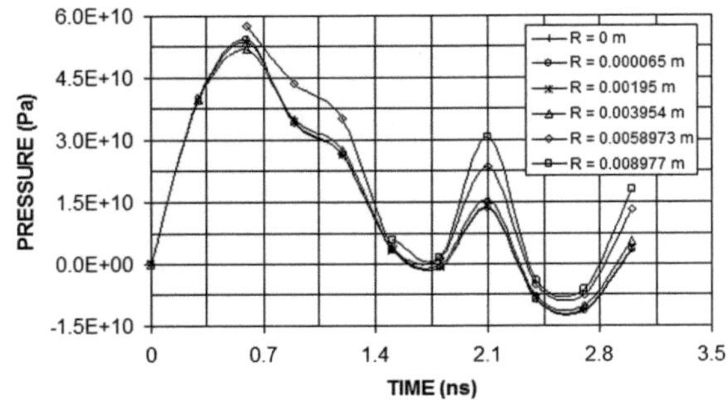

Fig. 5.14 Temporal variation of recoil pressure variation in the region of cavity exit at different radial locations

different heating durations. Since the size of the cavity is small, the value of maximum pressure does not alter significantly in the cavity. Some small changes in magnitude of pressure are observed in the cavity exit as compared to cavity wall. However, secondary peak of the recoil pressure is evident at the cavity exit, which is due to complex flow structure developed in the cavity. Moreover, the magnitude of secondary pressure peak is less than the initial pressure peak.

References

1. Yilbas, B.S., Kalyon, M.: Formulation of surface temperature for laser evaporative pulse heating: time exponentially decaying pulse case. Proc. Inst. Mech. Eng. Part C: J. Eng. Sci **216**, 289–300 (2002)
2. Yilbas, B.S., Mansour, S.B.: Laser evaporative heating of surface: Simulation of flow field in the laser produced cavity. J. Phys. D: Appl. Phys. **39**(17), 3863–3875 (2006)
3. Yilbas, B.S.: Analytical solution for time unsteady laser pulse heating of semi-infinite solid. Int. J. Mech. Sci. **39**(6), 671–682 (1997)
4. Abramowitz, M., Stegun, I.A.: Handbook of Mathematical Functions. Dover, New York (1965)
5. Shuja, S.Z., Yilbas, B.S., Budair, M.O.: Modeling of laser heating of solid substance including assisting gas impingement. Numer. Heat Transfer, Part A **33**, 315–339 (1998)
6. Naqavi, I.Z.: Conduction and non-conduction limited laser heating process—mathematical simulation. M.Sc. thesis, Mechanical Engineering Department, KFUPM (2001)
7. Patankar, S.V.: Numerical Heat Transfer. McGraw-Hill, New York (1980)

Chapter 6
Practical Applications of Laser Surface Treatment

Absract Laser processing of metallic surfaces funds wide application in industry. Some of these applications include laser shock processing, laser gas assisted nitriding, and laser induced particle injection at the surfaces. Laser shock hardening is a potential candidate for surface treatment of metallic parts. This is because the high power laser beam processing, which efficiently supplies energy directly to a small region of any material while resulting in localized treatment without bulk heating of the substrate. Hence, the properties of the bulk material remain unaffected. In this chapter, practical applications of laser treatment of metallic surfaces are presented. The morphological and metallurgical changes in the treated layer are discussed.

6.1 Introduction

Laser processing of metallic surfaces finds wide application in industry. Some of these applications include laser shock processing, laser gas assisted nitriding, and laser induced particle injection at the surfaces. Laser shock hardening is a potential candidate for surface treatment of metallic parts. This is because the high power laser beam processing, which efficiently supplies energy directly to a small region of any material while resulting in localized treatment without bulk heating of the substrate. Hence, the properties of the bulk material remain unaffected. However, heating of a small region results in a substantial temperature rise over a small distance in the substrate. This produces excessive thermal stresses in the region where the temperature gradient is extremely high; in some cases, formation of surface defects, such as cracks and surface fractures, which becomes unavoidable. In the case of short pulses, the recoil pressure generated due to surface ablation results in a plastic region formation in the surface vicinity. The plastic region possesses improved mechanical and tribological properties.

B. S. Yilbas and S. Z. Shuja, *Laser Surface Processing and Model Studies*, Materials Forming, Machining and Tribology, DOI: 10.1007/978-3-642-36629-1_6,

In the case of laser gas assisted nitriding, the metallic surface is heated by a laser beam and assisting nitrogen gas impinges onto the irradiated surface. Nitrogen diffuses and mixes with the molten metal forming a nitride layer in the surface vicinity. Depending on the laser output power intensity, laser scanning speed, nitrogen flow rate, and material properties, nitride depth changes within in 10–100 μm. However, from the application point of view, the end product quality is important. Therefore, the quality of the end product determines the applicability of the nitriding process. Increasing laser power intensity increases the depth of nitride layer; however, diffusion induced stresses can cause cracks/crack network formation within the nitride layer due to high cooling rates while lowering the end product quality. Consequently, laser gas assisted nitriding process is not straight forward, but requires investigation for the optimum nitriding conditions. However, the carbon coating, blended with TiC powders, at the surface prior to the laser treatment enables to form carbonitride compounds in the treated region. The treatment process involves with the initial preparation of the surface and the laser controlled melting of the prepared surface at nitrogen gas ambient. Since the process is complex in nature, the detailed investigation into the surface treatment process becomes essential to improve the surface properties and to reduce surface defects. Laser shock processing, laser gas assisted nitriding, and laser treatment of pre-prepared surface for improved properties are presented herein in line with the previous studies [1–3].

6.2 Laser Shock Processing

Laser shock processing is involved with ablation of the surface. In this case, incident laser beam is absorbed in the surface region of the substrate material. Once the amount of energy absorbed in this region exceeds the latent heat of evaporation of the substrate material, ablation initiates. Since the ablating front velocity is high, the recoil pressure is generated at the interface of the ablated front and the solid surface. Depending on the magnitude of recoil pressure, a stress wave propagating into the substrate material is developed. When the stress level exceeds the yielding limit of the substrate material, plastic deformation occurs. The formation of excessive arrays of dislocations, and the formation of other phases or twins in the plastic region causes hardening of the substrate material. Moreover, once the ablation diminishes the unloading elastic wave is generated, which propagates towards the solid bulk. Since the elastic wave propagates faster than the plastic wave and it meets with the plastic wave at some depth below the surface. The depth at which both waves met defines the plastic region of the substrate material. However, the use of Q-switched pulses ($\sim$150 ps) may have disadvantages over the nanosecond laser pulses. This is due to that the rapid relaxation of the plastic wave by release-waves (elastic waves) immediately after the pulse terminates. This decreases the propagation depth of the plastic wave inside the material, therefore, the desired effect is limited to a very shallow depth.

6.2.1 Heating, Recoil Pressure, and Wave Analysis

The analysis of shock processing covers the surface evaporation, recoil pressure formation, and elasto-plastic wave propagation. The mathematical analysis related to the shock processing is presented in line with the previous study [1]. The heat transfer equation including the evaporation due to a laser heating pulse can be written as:

$$k\frac{\partial^2 T(x,t)}{\partial x^2} + \rho CpV\frac{\partial T}{\partial x} + I_o(1-r_f)\exp(-\beta t)\delta\exp(-\delta x) = \frac{\partial}{\partial t}\left(\rho C_p T(x,t)\right) \tag{6.1}$$

where, k is the thermal conductivity, ρ is the density, δ is the absorption coefficient, Cp is the specific heat capacity, V is the recession velocity of the surface, which can be determined from the energy balance at the surface [4], i.e.:

$$V = \frac{I_o}{\rho[CpT + L_{ev}]} \tag{6.2}$$

An exponential time varying heating pulse is considered $\left(I_o(1-r_f)\exp(-\beta t)\right)$. The initial condition is:

$$\text{At}t = 0; \quad T(x,0) = 0$$

The boundary conditions are:

$$\text{Atthesurface;} \quad \left.\frac{\partial T(x,t)}{\partial x}\right|_{x=0} = \frac{\rho}{k}VL_{ev}$$

and

$$\text{At}x = \infty; \quad T(\infty,t) = 0$$

where L_{ev} is the latent heat of evaporation.

The recoil pressure can be calculated after considering the rate of momentum change across the receding surface [5]. The rate of momentum, which occurs across the receding surface is:

$$A.P_r = \dot{m}V_{ev} \tag{6.3}$$

where A is the area of irradiated surface, P_r is the recoil pressure, $\dot{m}$ is the rate of mass flow and V_{ev} is the evaporating front velocity, which can be formulated from the Maxwell's law for the velocity distribution of molecules [4]:

$$V_{ev} = \sqrt{\frac{k_B T_s}{2\pi m}}\exp\left(-\frac{mL_{ev}(T)}{k_B T_s}\right) \tag{6.4}$$

where m is the mass of an atom.

Equation 6.3 can be rearranged and the recoil pressure can be written as:

$$P_r = \frac{\dot{m}}{A} V_{ev} \tag{6.5}$$

When the irradiated energy is absorbed by the substrate material some part of absorbed energy is consumed as heat transferred to the solid substrate due to conduction and the other part is used for the evaporation of the surface. Consequently, energy dissipated for evaporation per unit time per unit area results in releasing of molecules from the surface and acceleration of vapor molecules to V_{ev} velocity, i.e. energy dissipated per unit time due to evaporation is the sum of latent heat of evaporation and the kinetic energy of the evaporated molecules per unit time. This results in:

$$I_{ev} = \frac{\dot{m}\left(\frac{1}{2}V_{ev}^2 + L_{ev}\right)}{A} \tag{6.6}$$

Substitution of $\frac{\dot{m}}{A}$ in the above equation yields:

$$I_{ev} = \frac{P_r}{V_{ev}}\left(\frac{1}{2}V_{ev}^2 + L_{ev}\right) \tag{6.7}$$

Therefore, the recoil pressure can be written as:

$$P_r = \frac{I_{ev}V_{ev}}{\left(\frac{1}{2}V_{ev}^2 + L_{ev}\right)} \tag{6.8}$$

The recoil pressure can be computed from Eq. 6.8.

The laser power intensity dissipated due to evaporation is:

$$I_{ev} \simeq I_o - \dot{q}_{solid} \tag{6.9}$$

where $\dot{q}_{solid}$ is the energy dissipated in the solid phase. It should be noted that the energy dissipated due to melting is considerably small as compared to energy dissipated during evaporation.

The vapor pressure predicted from 6.8 is validated using the equation given in the previous study [6]; therefore:

$$P_s = 1.82 \times 10^{-3} \frac{\sqrt{CpT_{ev}}I_o}{[Cp(T_{ev} - T_o) + L_v]} \tag{6.10}$$

Equation 6.10 is expressed in units of Bar. When comparing the results of 6.8 and 6.10, it found that they are in good agreement.

When the free surface of the solid substrate is subjected to an impact force due to a suddenly generated pressure, an elastic–plastic wave, which propagates into the solid is generated. The wave amplitude composes of two parts, which are the amplitude greater and less than the yield stress. The amplitude of the wave greater than yield stress results in the plastic deformation inside the substrate. As soon as the impact load released from the surface, an loading wave is initiated propagating through the substrate. The point, where the unloading wave intersects the elastic–plastic wave, determines the depth of the plastic region formed in the substrate. The mathematical analysis governing the wave propagation is given in [7]; therefore, only the basic equations are given below. The local momentum balance for homogeneous mass of a solid can be written as [7]:

$$\rho \frac{\partial u}{\partial t} = \frac{\partial \sigma}{\partial x} \tag{6.11}$$

where u is the velocity in the x-direction, σ is the normal stress, ρ is the density of the substrate, and t is the time. When the displacement is denoted by U and strain by ε, the following relations can be written:

$$u = \frac{\partial U}{\partial t} \text{ and } \varepsilon = \frac{\partial U}{\partial x} \tag{6.12}$$

The integrability condition of the substrate results in:

$$\frac{\partial \varepsilon}{\partial t} = \frac{\partial u}{\partial x} \tag{6.13}$$

For a linear elastic material, Hooke's law gives:

$$\sigma = E\varepsilon \tag{6.14}$$

where E is a Young's modulus. Substitution of σ into Eq. 6.13 yields:

$$\frac{\partial w}{\partial t} = A \frac{\partial w}{\partial x} \tag{6.15}$$

where

$$w = \begin{pmatrix} \rho u \\ \varepsilon \end{pmatrix} \text{ and } A = \begin{pmatrix} 0 & E \\ \frac{1}{\rho} & 0 \end{pmatrix} \tag{6.16}$$

Therefore:

$$\frac{\partial w}{\partial t} = \frac{\partial f}{\partial x} \tag{6.17}$$

where $f = (\sigma, u)^T$.

In the case of elastic–plastic material, the loading history of the boundary and work-hardening of the substrate should be considered. To account for the elastic–plastic condition, the stress–strain relation can be written as [7]:

$$d\varepsilon = \frac{1 + h}{E} d\sigma \tag{6.18}$$

where $h = h(\kappa)$ is defined as the plastic factor, which has the following conditions:

$$h = 0 \text{ at } |\sigma + d\sigma| \leq \kappa$$
$$\text{and } h = \frac{E}{E_p(\kappa)} - 1 \text{ at } |\sigma + d\sigma| > \kappa \tag{6.19}$$

Substitution of $d\varepsilon$ into Eq. 6.17 yields:

$$A\frac{\partial f}{\partial t} = \frac{\partial f}{\partial x} \tag{6.20}$$

where:

$$A = \begin{pmatrix} 0 & \rho \\ \frac{1+h}{E} & 0 \end{pmatrix} \text{ and } \rho = \begin{pmatrix} \sigma \\ u \end{pmatrix} \tag{6.21}$$

6.2.2 Method of Solutions

To discretize the governing equations, a finite difference approach is introduced. The details of the numerical scheme are given in [8]. To compute the temperature field and relative positions of solid–vapor interface, an explicit scheme (a time marching scheme) is used, i.e. using the initial conditions, the temperature in the whole domain is calculated for following time steps with the respective conditions. This process is repeated for the next time step and $V \neq 0$ is introduced where ever temperature reaches to the boiling temperatures of the substrate material. The stability criteria due to time increment are considered for a stable solution; therefore, the time increment is limited by:

$$\left(\frac{2k}{\rho c_p \Delta z^2}\right) \Delta t \leq 1 \tag{6.22}$$

Time increment satisfying the stability criteria for the finest grid is calculated as 10^{-13} s. A computer program based on explicit scheme is developed to compute the temperature field.

Table 6.1 Material properties used in the analysis

	$\delta \times 10^7$ (1/m)	$\alpha \times 10^{-4}$ (m^2/s)	Cp (J/kgK)	λ (m)	k (W/mK)	L_m (KJ/kmol)	L_v (KJ/kmol)	T_m (K)	T_v (K)
Stainless steel	6.16	0.227	460	10^{-9}	80.3	13.8	349.5	1,892	3,166
Mild steel	6.16	0.231	447	10^{-9}	80.2	13.8	349.5	1,810	3,166

Table 6.2 Mechanical properties of materials used in the analysis

Material	P (GPa)	σ_y (MPa)	E (GPa)	ρg (kN/m^3)	α	τ_t (ns)	ν	L_p (mm)
Stainless steel	1	200	207	70.69	2	18	0.3	0.470
Mild steel	1	165	206	77.20	2	18	0.29	0.480

The material properties used in the simulations are given in Table 6.1.

The elastic–plastic boundaries in the time and space (x, t) domain are established by obtaining the first plastic $\left(\frac{\kappa}{\kappa_o}\right)$ wave and the elastic unloading wave. A semi-infinite body which obeys a linearly elastic, power-law work-hardening plastic stress–strain relationship is considered, which may be written as:

$$\frac{1}{E(\kappa)} = \frac{\alpha}{E}\left(\frac{\kappa}{\kappa_o}\right)^{\alpha-1} \tag{6.23}$$

The solid body is assumed to be at rest at $t<0$, which in turn yields the initial conditions for u and σ are zero. The boundary condition at $x=0$ includes a rectangular impact load, i.e.:

$$\begin{aligned} \sigma(0,\, t) &= P_r \text{ (1 GPa) for } 0 \le t \le t_p \text{ ns} \\ \sigma(0,\, t) &= P_r \text{ (0 GPa) for } t_p \text{ ns} \le t \le \infty \end{aligned} \tag{6.24}$$

where t_p is the twice of the pulse length $\left(t_p = 2 \times 9\,\text{ns}\right)$. It should be noted that due to the overlay placed at the workpiece surface in the experiment, t_p is set twice the pulse length as suggested in the previous study [6]. A computer program is developed to solve the governing stress wave equation numerically using the Godunov scheme [9]. The numerical procedure consists of three steps: (1) Solving the Riemann problem to obtain the solution for the fan-shaped simple wave regions on both sides of the interface, (2) Calculating $\bar{f}_{j+1/2}$, and (3) updating w_j^{n+1} for u and σ. The mechanical properties used in the simulations are given in Table 6.2.

6.2.3 Experimental

To irradiate the workpiece surfaces a Nd:YAG laser was used. Laser delivered a nominal output energy of 450 mJ within a pulse length of about 8 ns. The repetition rate of the laser pulses was 10 Hz. In order to increase the laser output pulse intensity a focusing lens with a nominal focal length of 100 mm was employed. The diameter of the heated spot was kept at about 2.5 mm at the workpiece surface. 15 consecutive pulses were fired during the shock processing of the surface. An overlay of 0.1 mm thickness was mounted on the workpiece surface to enhance the recoil pressure developed at the interface. It should be noted that using an overlay increased substantially the magnitude of the recoil pressure.

Table 6.3 Elemental composition of the workpiece prior and after laser shock treatment (wt %)

		C	Mn	Si	Cr	Ni	P	S	Fe
Mild steel	Before	0.07	0.2	0.1	–	–	–	–	Balance
	After	–	0.02	0.1	–	–	–	–	Balance
Stainless steel	Before	–	2.5	0.6	18.1	6.5	0.1	0.1	Balance
	After	–	1.3	0.5	21.8	6.0	–	–	Balance

Moreover, during the ablation, a local melting could produce high residual stress levels in the surface vicinity of the workpieces, i.e. the confined regime develops.

A commercial stainless steel 316 was used. The chemical composition of the workpieces is given in Table 6.3 and they were cut in rectangular shape with $12 \times 30 \times 2\ mm^3$ (width $\times$ length $\times$ thickness). The workpieces were cleaned ultrasonically before the experiments. The cross-sections of the workpieces were obtained and etched after the laser treatment process.

SEM (JEOL JSM-T300) was used for microstructural analysis while microhardness across the workpiece surfaces was measured with a Vickers microhardness equipment (Puhlar Micromet 3). During the microhardness tests a load of 300 gr was used. TEM (JEOL 2527-TEM SCAN) was utilized to obtain the TEM micrographs for determining the dislocation densities in the laser shocked zone.

6.3 Laser Gas Assisted Nitriding

Surface modification through synthesis of nitrides in the surface region of metallic alloys provides superior tribological properties. In industry, many nitriding techniques are used for metallic substrates and some these techniques include reactive magnetic sputtering, plasma nitriding, ion implantation, salt bath nitriding, etc. However, laser can offer significant advantages over the conventional nitriding processes due to the precision of operation, short processing duration, and local processing without changing the bulk properties of the substrate material. Laser gas-assisted nitriding of steel surfaces is possible through generating the nitride species in the irradiated region. The laser nitriding provides hard surfaces, which resist wear, erosion, and corrosion. On the other hand, laser nitriding process is involved with high temperature heating and high cooling rates. This, in turn, results in high levels of thermal stresses in the heated region. The residual stress in the heated region remains high after the completion of the heating process, which limits the application of the laser nitriding process. Therefore, further developments into laser gas assisted nitriding with low residual stress levels in the nitrided layer become essential. In this section, laser nitrogen assisted nitriding of steel surface and the residual stress developed in the laser treated layer are presented in the light of the previous study [2].

6.3.1 Thermal and Residual Stress Analysis

The transient diffusion equation based on the Fourier heating model can be written in the Cartesian coordinates as:

$$\rho \frac{\partial (CpT)}{\partial t} = (\nabla (k \nabla T)) + \rho U \frac{\partial (CpT)}{\partial x} + S_o \tag{6.25}$$

where ρ is the density, Cp is the specific heat capacity, k is the thermal conductivity, U is the laser scanning speed, S_o is the source term, which is:

$$S_o = I_o \delta (1 - r_f) e^{\left(-\frac{x^2 + z^2}{a^2}\right)} \tag{6.26}$$

I_o is laser peak intensity, δ is the absorption depth, r_f is the surface reflectivity, a is the Gaussian parameter and x, and z are the axes (Fig. 6.1), ρ is the density, C_p is the specific heat capacity, and k is the thermal conductivity. The laser beam axis is located along the y-axis (Fig. 6.1).

At the free surfaces of the nitrided specimen (Fig. 6.1), the convective boundary is assumed, except at the irradiated surface at which convective and radiative boundary condition is considered. Therefore, the corresponding boundary condition is:

$$\frac{\partial T}{\partial x} = \frac{h}{k}(T_s - T_{amb}) : \frac{\partial T}{\partial y} = \frac{h}{k}(T_s - T_{amb}) + \frac{\varepsilon \sigma}{k}\left(T_s^4 - T_{amb}^4\right) : \frac{\partial T}{\partial z} = \frac{h}{k}(T_s - T_{amb})$$

where h is the heat transfer coefficient due to natural convection, and T_s and T_{amb} are the surface and ambient temperatures, respectively, ε is the emissivity ($\varepsilon = 0.9$ is considered), σ is the Stefan-Boltzmann constant ($\sigma = 5.67 \times 10^{-8}$ W/m^2 K). At far away boundary (at edges of the solution domain) constant temperature boundary is assumed (T = 293 K), i.e.

$$x = \infty; \quad y = \infty; \quad z = \infty \rightarrow \mathrm{T} = 20^\circ \mathrm{C}$$

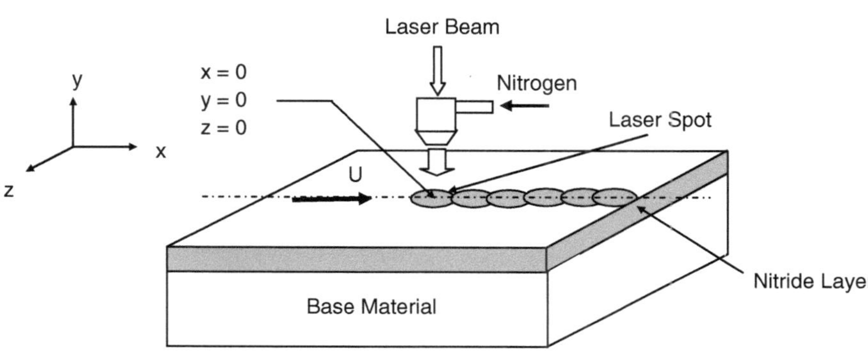

Fig. 6.1 Schematic view of laser gas assisted nitriding and coordinate system

Table 6.4 Thermal properties used in the simulations. Density is 7,700 kg/m^3

Temp (K)	273	373	473	573	673	773	873	973	1,073	1,273	1,473
k (W/mK)	51.9	50.7	48.2	45.6	41.9	38.1	33.9	30.1	24.7	26.8	29.7
Cp (J/kgK)	486	486	515	528	548	586	649	708	770	624	548

Initially (prior to laser treatment), the substrate material is assumed to be at constant ambient temperature, i.e. $T = T_{amb}$, which is considered as constant ($T_{amb} = 293$ K).

Equation 6.25 is solved numerically with the appropriate boundary conditions to predict the temperature field. Table 6.4 gives the data used in the simulations.

The nitrogen diffusion during the laser processing can be formulated using the diffusion equation;

$$\rho \frac{\partial (C)}{\partial t} = (\nabla (D \nabla C)) \tag{6.27}$$

where C is the concentration and D is the temperature dependent diffusion coefficient. It should be noted that the data obtained from the previous study was used for the temperature dependent diffusion coefficient [10].

The initial and the boundary conditions for Eq. (6.27) are:

At time $t = 0$ (prior to the laser treatment process), nitrogen concentration (C) in the substrate material is assumed to be zero at all locations, i.e. $C = 0$.

At time $t > 0$ (during the laser processing):

At the workpiece surface, the nitrogen concentration gradient is set to constant. This is because of the assumption of constant nitrogen concentration above the surface of the workpiece during the laser processing, i.e.:

$$\frac{\partial (C)}{\partial y} = \text{Constant} \tag{6.28}$$

At a depth well below the surface, the nitrogen concentration is assumed to be zero. This is because of the assumption that at a depth of infinity, nitrogen diffusion ceases, i.e.:

$$C = 0 \tag{6.29}$$

Equation 6.27 with the appropriate boundary conditions was used to predict the nitrogen concentration in the substrate material.

Table 6.5 The data used in Eq. 6.20

E_c (GPa)	E_s (GPa)	$t_s \times 10^{-3}$ (m)	$t_c \times 10^{-6}$ (m)	$\alpha_c \times 10^{-6}$ (1/K)	$\alpha_s \times 10^{-6}$ (1/K)	T_R (K)	T_f (K)
140	193	2	60	15.5	16.2	300	2,300

The analytical expression for the residual stress was developed previously is used [11]. The previous expression is based on the thermal expansion mismatched between the coating and the base material. Therefore, the residual stress is:

$$\sigma = \frac{\left[E_c\left(T_f - T_R\right)\left(\alpha_c - \alpha_s\right)\right]}{1 + 2\left(\frac{E_c t_c}{E_s t_s}\right)} \tag{6.30}$$

where E_c and E_s are the elastic modules of the coating and the base material, t_c and t_s are the coating and base material thicknesses, T_f is the maximum temperature during laser heating and T_R is the room temperature after the cooling period is over, α_c and α_s are the thermal expansion coefficient of the coating and the base material, respectively. Table 6.5 gives the data used for the residual stress calculation using Eq. 6.30.

6.3.2 Experimental and Measurement of Young Modulus and Fracture Toughness

The CO_2 laser (LC-ALPHAIII) delivering nominal output power of 2 kW at pulse mode with different frequencies is used to irradiate the workpiece surface. The nominal focal length of the focusing lens is 127 mm. Nitrogen assisting gas emerging from the conical nozzle and co-axially with the laser beam is used. The nitriding conditions are given in Table 6.6. The workpiece accommodated is steel at 2 mm in thicknesses.

Material characterization of the laser nitrided surfaces is carried out using SEM, XRD and XPS. Jeol 6460 electron microscopy is used for SEM examinations and Broker D8 Advanced having MoKα radiation is used for XRD analysis. A typical setting of XRD was 40 kV and 30 mA. Nitrogen concentration and depth profiling in the surface region is performed through X-ray photoelectron spectroscopy (XPS). The XPS measurements were carried out in an electron spectrometer (VG ESCALAB MKII) equipped with a computer controlled ion gun (Type EXO5). The X-ray source was CuKα radiation. N1s, O1s Fe2p3/2, and Cr2p3/2 lines were recorded before and after successive etching cycles using Ar+ ions of 4 keV energy. The etched area was 6 × 6 mm. The target current was 0.5 μA and the pressure in the chamber during sputtering was 10^{-7} mBar. The C1s line of

Table 6.6 Laser assisted nitriding conditions

Feed rate (mm/min)	Power (W)	Frequency (Hz)	Nozzle gap (mm)	Nozzle diameter (mm)	Focus setting (mm)	N_2 pressure (kPa)
1,000	70 and 110	1,500	1.5	1.5	127	600

284.6 eV binding energy was used as a reference to correct the binding energies for the charge shift.

Microphotonics digital microhardness tester (MP-100TC) was used to obtain microhardness across the depth of the nitride layer. The load used during the measurements was kept as 100 gf (0.9807 N). The standard test method for Vickers indentation hardness of advanced ceramics (ASTM C1327-99) was adopted. Microhardness at the surface was measured at the free surface of the workpiece after metallurgical preparation. The workpiece cross-section was also prepared for the microhardness measurements. The measurements were repeated three times at each location.

The elastic response of the surface when subjected to indention test needs to be examined through which the Young's modulus can be determined. The Young's modules can be formulated as [12]:

$$E^* = \left(\frac{9}{16}\right)^{0.5} P.h^{-1.5}.R^{-0.5} \tag{6.31}$$

where P is the applied load, h is the elastic penetration of the indenter, and R is the indenter radius. The true modulus of elasticity can be determined using the indenter properties [13], i.e.:

$$E = \frac{1 - \nu^2}{\frac{1}{E^*} - \frac{\left(1-\nu_i^2\right)}{E_i}} \tag{6.32}$$

where E_i and ν_i are the Young's modulus and Poisson's ratio of indenter, respectively. In the calculations E_i = 1,141 GPa and $\nu_i = 0.07$ (diamond indenter) [14], and $\nu = 0.25$ are taken.

The fracture toughness of the surface is measured using the indenter test data for microhardness (Vickers) and crack inhibiting. In this case, microhardness in HV and the crack length generated due to indentation at the surface are measured. The fracture toughness (K) is calculated using the equation given below [15]:

$$K = 0.016\left(\frac{E}{H}\right)^{0.5}.P.C^{-1.5} \tag{6.33}$$

where H is the hardness (Vickers), P is the applied load on indenter, C is the crack length. The data used for the calculations are given in Table 6.7. Moreover, in

Table 6.7 Data obtained after indenting tests and used to determine the elastic modulus and fracture toughness of the laser nitrided surface

	H (HV)	*P* (N)	*C* (μm)	*R* (μm)	*h* (μm)
As received	360 (3.53 GPa))	20	8.5	60	5
Laser nitrided	820 (8.04 GPa)	20	35	60	6

order to visualize the cracks formed around the indention mark, top surface of coating is grinded slightly.

6.4 Laser Surface Treatment of Pre-prepared Alloy

Ti-6Al-4V alloy finds applications in industry because of its high toughness to mass ratio; however, the poor surface properties of the alloy limit its applications in practice. One of the methods to improve the surface properties of the alloy is to form a hard layer at the surface. In this case, the laser surface treatment process offers advantages over the conventional methods. Some of these advantages include precision of operation, control processing, short process duration, and the low cost. Moreover, the carbon coating, blended with TiC powders, at the surface prior to the laser treatment enables to form carbonitride compounds in the treated region. The treatment process involves with the initial preparation of the surface and the laser controlled melting of the prepared surface at nitrogen gas ambient. Since the process is complex in nature, the thermal modeling is involved errors because of the assumptions made for the simplifications. Therefore, the experimental investigation and findings are presented herein in the light of the previous study [3].

The CO_2 laser (LC-ALPHAIII) delivering nominal output power of 2 kW at pulse mode with different frequencies is used to irradiate the workpiece surface. The nominal focal length of the focusing lens is 127 mm. Nitrogen assisting gas emerging from the conical nozzle and co-axially with the laser beam is used. The nitriding conditions are given in Table 6.8.

The workpiece accommodated is Ti-6Al-4V alloy at 3 mm in thicknesses ($30 \times 40 \times 3$ mm^3: width × length × thickness). The water soluble phenolic resin (thermoset viny-phenolic resin) was applied to the workpiece surfaces and kept at 8 bar argon pressure and 175 °C for 2 h in the furnace. After ensuring the conversion of the phenolic matrix into carbon layer through monitoring the density, TiC particles were inserted mechanically into the carbon layer formed at the workpiece surfaces. The averaged size of the TiC particles was in the order of 6 μm and the volume fraction of TiC particles inserted into the carbon film was about 20 %.

Material characterization of the laser nitrided surfaces was carried out using SEM, EDS, and XRD. Jeol 6,460 electron microscopy is used for SEM examinations and Bruker D8 Advanced having MoKα radiation was used for XRD analysis. A typical setting of XRD was 40 kV and 30 mA. The parabolically-

Table 6.8 Laser gas assisted nitriding conditions

Feed rate (mm/min)	Power (W)	Frequency (Hz)	Nozzle gap (mm)	Nozzle diameter (mm)	Focus setting (mm)	N_2 pressure (kPa)
800	80	1,500	1.5	1.5	127	600

shaped Göbel Mirror was used in Bruker D8 Advanced, which provided the highly-parallel x-ray beams. The parallel beams from it removed the traditional limitations of the Bragg–Brentano geometry for powder diffraction and thin film investigations. Microphotonics digital microhardness tester (MP-100TC) was used to obtain microhardness at the surface of the nitride layer. The standard test method for Vickers indentation hardness of advanced ceramics (ASTM C1327-99) was adopted. The measurements were repeated three times at each location.

6.5 Discussion

The practical applications of laser surface treatment are important and three treatment techniques are incorporated in this chapter despite the fact that there are other techniques used in laser surface treatment of engineering surfaces. The findings of the techniques incorporated are discussed according to the sub-headings in line with the previous studies [1–3].

6.5.1 Laser Shock Processing of Steel Surface

Laser ablation of stainless steel and laser shock hardening of the surface are presented in line with the previous study [1]. The recoil pressure generated at the surface during the ablation is determined numerically. The loading force generated due to recoil pressure is considered as a force generating a plastic zone in the surface region of the substrate material.

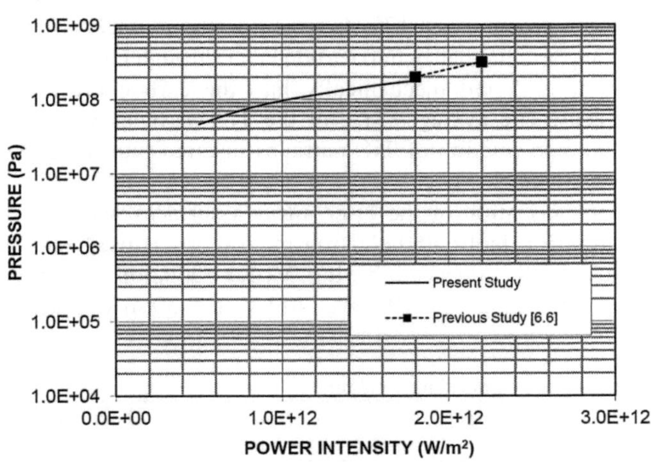

Fig. 6.2 Recoil pressures with power intensity

Figure 6.2 shows the recoil pressure generated at the surface. The recoil pressure increases considerably with increasing power intensity; in which case, high temperature at the surface occurs with increasing power intensity. This, in turn, accelerates the vapor front as well as recession velocities of the surface. Consequently, considerably high magnitude of recoil pressure is generated due to the momentum change in the direction of flow. When comparing the experimental values with the present predictions, both results are in agreement.

Figure 6.3 shows the stresses developed inside substrate material due to propagation of the plastic wave generated by the recoil pressure at the workpiece surface. The magnitude of stress attains high values in the workpiece surface vicinity in the early heating period and the stress due to plastic wave propagation is compression in nature. Moreover, once the evaporation ceases, recoil pressure reduces to zero and unloading wave with a higher speed than plastic wave speed propagates towards the substrate material. The unloading wave catches the plastic wave and reduces the plastic wave amplitude significantly. Consequently, plastic wave terminates at the point where plastic and unloading elastic waves meet. The time where these two waves interact determines the time taken for plastic region formation. In the present case, it is in the order of 1.2×10^{-7} s and the corresponding depth of plastic region is in the order of 400 μm.

Figure 6.4 shows the top view of the laser treated workpiece surface. The size of the surface ablated is in the order of 2.5 mm. The shallow cavity is formed at the surface. The laser evaporated and shock regions can be identified from the view of the surface. The formation of waviness and small recession of the surface are due to evaporation of the surface and the plastic deformation in the surface vicinity. Moreover, non-existence of the liquid layer at the surface also indicates that considerable portion of the incident laser energy is transferred to surface ablation process. The excessive compressive stress and thermal effect due to

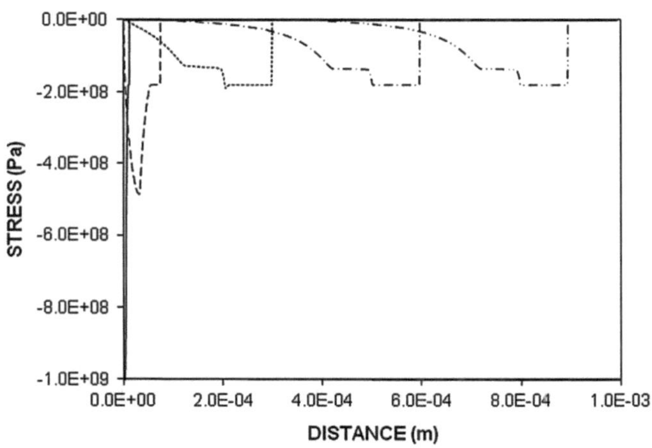

Fig. 6.3 Stress wave propagation for stainless steel with pressure = 0.8 GPa, pulse length = 8.0 ns and $dt = 0.05$ ns

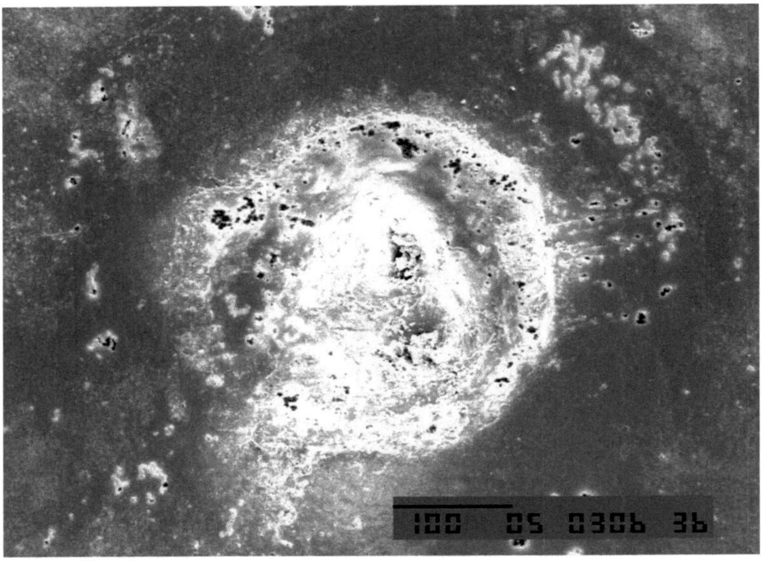

Fig. 6.4 SEM micrograph of laser shock processed surface

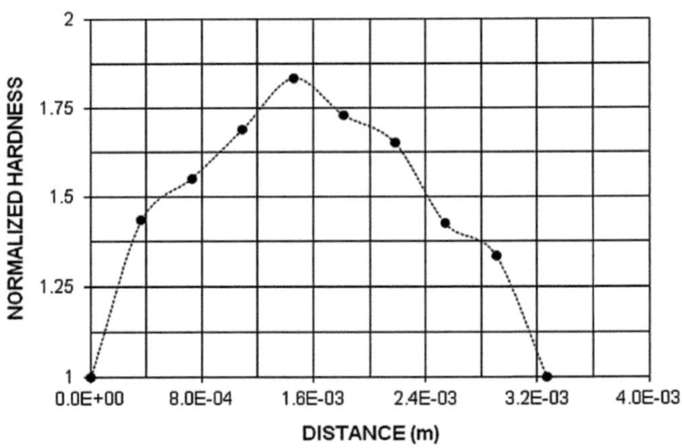

Fig. 6.5 Normalized microhardness distribution across the irradiated spot

confined regime generate a plastic zone in the surface region, as indicated in the previous study [6].

Figure 6.5 shows the microhardness distribution across the heated surface. Microhardness attains almost 1.8 times of the base substrate hardness in the central region of irradiated spot. The decay of the hardness across the heated spot towards its edge is because of the power intensity distribution at the surface, i.e. power intensity reduces from heated spot center to its edge. The increase in hardness is associated with the dislocation generated in the surface region of the irradiated spot during the heating process. The dislocation density as measured from transmission electron microscope (TEM) image is in the order of 2–8 × 10^{11} cm^{-2}. This, in turn, results the hardness of the treated region 1.8 times of the base metal. In this case, the relation between the hardness and dislocation density, which was developed previously [16], is used, since the dislocation density is known. The equation for the hardness due to dislocation is:

$$H = 145 + 2.1 \times 10^{-4}\sqrt{P} \tag{6.34}$$

where 145 is the hardness of the base material, 2.1× 10^{-4} is constant for steel, and P is the dislocation density. Therefore, substituting the dislocation density in the above equation, the hardness values in the ablated region is in the order of 250 HV is resulted, which is in good agreement with the measurements (Fig. 6.5).

6.5.2 Laser Gas Assisted Nitriding of Steel Surfaces

Laser gas-assisted nitriding of steel surface is carried out. Temperature field and nitrogen concentration in the substrate material are stress predicted numerically. The metallurgical changes and residual stress formed in the laser irradiated region are examined using scanning electron microscope (SEM), energy dispersive spectroscopy (EDS), X-ray-diffraction (XRD), X-ray Spectroscopy (XPS) and indentation tests. The findings are presented in line with the previous study [2].

Figure 6.6 shows temperature distribution in the substrate material at the end of the heating pulse. Temperature decays gradually in the surface region and it decreases sharply in the vicinity of the surface region. The gradual decay of temperature is because of the absorption of the incident beam in the surface region; in which case, internal energy gain of the substrate material increases significantly.

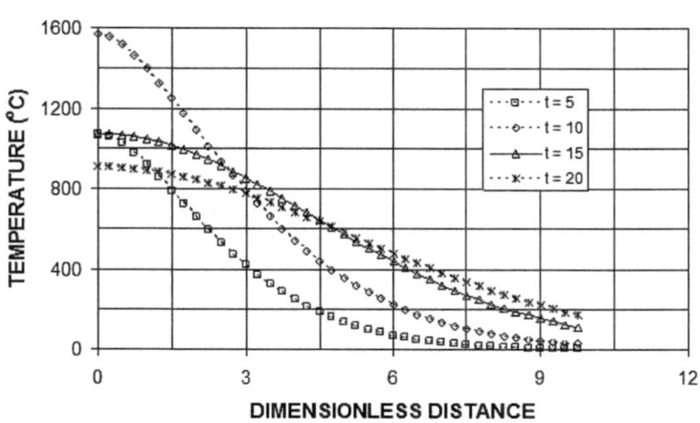

Fig. 6.6 Temperature distribution inside the substrate material for different time periods

Table 6.9 The residual stress predicted, measured using the XRD technique, and calculated from the analytical expression

Predictions (MPa)	XRD technique (MPa)	Analytical expression [11] (MPa)
230	210	190

The residual stress predicted and measured is tensile, which is similar to that reported in the literature [17]

Table 6.10 Elastic modulus of laser nitrided layer and fracture toughness of the as received and laser nitrided surfaces

Elastic modulus laser nitride (GPa)	Fracture toughness laser nitride surface ($MPa/m^{1/2}$)	Fracture toughness un-treated surface ($MPa/m^{1/2}$)
140	6.44×10^6	9.36×10^7

However, in the region next to the surface vicinity, the amount of energy absorbed by the substrate material reduces. This is because of the amount of laser power absorbed, which reduces with increasing depth from the surface (Lambert's law). Moreover, temperature gradient in this region remains high due to sharp decay of temperature. This, in turn, enhances the rate of heat diffusion from the surface region to the solid bulk. Consequently, temperature rises in the region next to the surface vicinity is governed by the heat diffusion rather than the laser power absorbed by the substrate material.

The comparison of the residual stress obtained from the experiment using both the XRD technique and the indentation tests as well as the analytical method is given in Table 6.9. It should be noted that the residual stresses are tensile. Moreover, it can be observed that all the results are in reasonable quantitative agreement. However, some small discrepancies between the residual stress findings are because of the following reasons: (1) the substrate material is considered to be homogeneous and the stress field is governed by the thermal affect only while in the actual case material is not an homogeneous medium and metallurgical changes contributed to the residual stress formation, and (2) nitrogen diffusion results in stress field in the diffusion zone; therefore, it is contribution to residual stress becomes important, particularly in the region where the nitrogen concentration is high. Nevertheless, the predictions of the residual stress give insight into the stress concentration and distribution in the laser irradiated region with a reasonable accuracy.

The elastic modulus of the laser nitrided region and the fracture toughness measured for the laser nitrided surface are given in Table 6.10. The data for the as received workpieces are also given for comparison. Elastic modulus of the nitrided region is similar to that presented in the literature [14]. The fracture toughness of the surface reduces considerably after the laser treatment process due to the surface brittleness resulted from the nitriding process. This decrease is significant and the data obtained is similar to that obtained in the previous study [18].

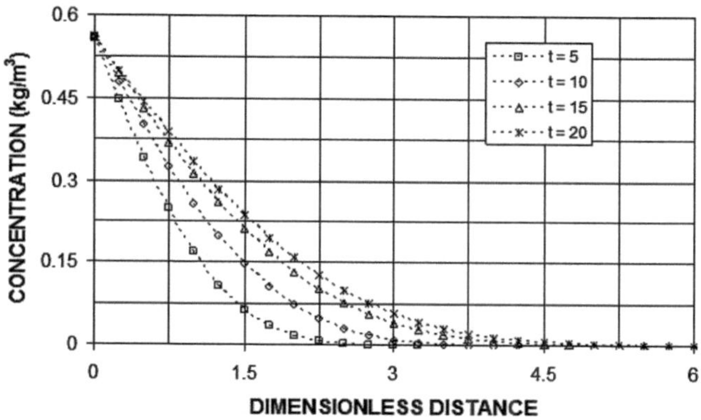

Fig. 6.7 Concentration distribution inside the substrate material for different time periods

Table 6.11 XPS data for the laser nitrided workpiece

	As prepared surface	After Ar etching Ar+, 4 keV, 5 min, target current 1.5 μA
Fe (% at)	49	87
N (% at)	51	13

Figure 6.7 shows normalized nitrogen concentration in the substrate material. The predictions are obtained after considering temperature dependent diffusion coefficient [10]. The nitrogen concentration is high in the surface vicinity and sharp decay is observed as the depth from the surface towards the solid bulk increases. This is attributed to: (1) temperature decay is sharp in this region and the

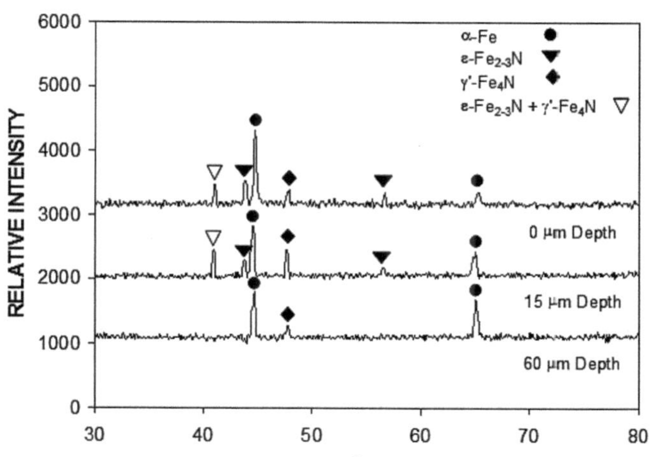

Fig. 6.8 XRD results for laser nitrided workpiece at different depths

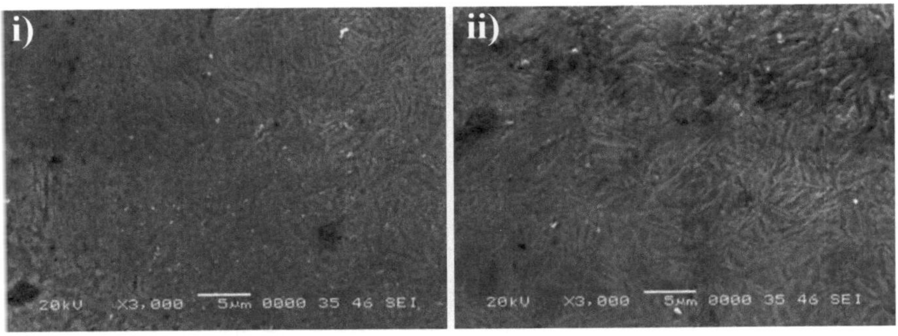

Fig. 6.9 SEM micrographs of two surfaces: **a** at the free surface (y = 0 m) and **b** at a depth 15 μm below the free surface of the nitrided layer (y = 15 μm)

diffusion coefficient reduces significantly, and (2) nitrogen diffusion is governed by the diffusion equation, and the gradient of species concentration is the main driving factor for the diffusion process, i.e., sharp decay of gradient suppresses the nitrogen diffusion into the substrate material. High nitrogen concentration in the surface region indicates that the formation of ε–$Fe_{2-3}N$ in this region is most likely. This situation is also observed from the XPS measurements as shown in Table 6.11. However, this depth is limited with the shallow layer in the surface vicinity. Consequently, sharp reduction in the nitrogen concentration with increasing depth reveals that formation of γ'-Fe_2N is likely just below the surface.

Figure 6.8 shows XRD results obtained at different depths of the nitrided layer. The different depths are obtained through a grinding of the workpiece surface at 15 and 60 μm depths below the surface. It should be noted that the depth of grinded surface was measured through electron microscopy. It is observed that in the surface region combination of ε–Fe_2N and ε–Fe_3N occurs. The high nitrogen content in the surface vicinity is because of the nitrogen diffusion which is governed by the concentration flux, i.e., high concentration flux is responsible for high nitrogen content at the surface (Fig. 6.7). However, when examining the next layer (15 μm below the surface), ε–$Fe_{2-3}N$ replaces with mixtures of ε–$Fe_{2-3}N$ and γ'-Fe_4N precipitates. It should be noted that the rate of nitrogen diffusion is expected to be high at the grain boundaries [10]. As the laser gas-assisted nitriding progresses, the nitrogen diffusion takes place from the grain boundaries to the grains. High nitrogen concentration results in very fine structure, which can be observed from SEM micrograph (Fig. 6.9). However, as the nitride species coarsens, the elongated and large structures are observed (Fig. 6.9). As the depth removed below the surface increases towards the solid bulk (depth of about 60 μm) ε–$Fe_{2-3}N$ phases ceases and the intensity of γ'-Fe_4N reduces, i.e., it is only γ'-Fe_4N phase becomes notable in the XRD graph. In addition, the large structures with some small nitride compounds are dominant at this depth.

Figure 6.10 shows SEM micrographs of the top surface and cross-section of the laser nitrided workpiece. It is evident from both micrographs that nitrided surface

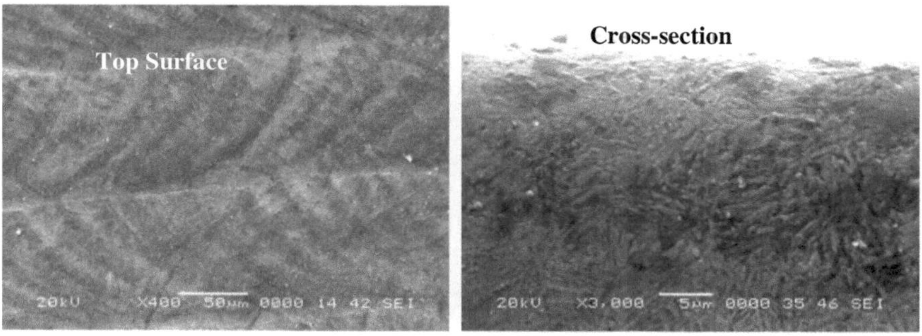

Fig. 6.10 SEM micrographs of the top surface and the cross-section of the laser nitrided layer

is free from local cavitations and the cracks. However, the overlaying pattern of the laser melted zone is evident. The regular appearance of the melt pattern is associated with the laser scanning speed, provided that at higher laser scanning speed the regular pattern width reduces slightly because of the localized melt at the surface and minimized thermal agitation at high scanning speeds. Moreover, in all scanning speeds, the variation in the surface roughness is minimal. It is evident from the SEM micrograph (Fig. 6.10) that the fine feathery-like fine structures are the representative of the nitride-rich compounds $Fe_{2-3}N$ at the surface. However, the depth of nitride-rich zone is significantly shallow as observed from the cross-sectional views. This region is brittle in structure and associated with the white zone as also observed in the previous study [10]. The depth of nitride layer extends slightly more than 60 μm below the surface. However, the high power laser irradiation initiated the evaporation at the surface while forming the cavity and increasing the surface roughness significantly. Consequently, the laser power was set to avoid the excessive surface evaporation during the nitriding experiments. The regular pattern appearing in the nitrided zone follows the base melt tracks. Moreover, almost uniform with constant depth of nitride layer formation is evident from the figure.

6.5.3 Laser Surface Treatment of Pre-prepared Ti Alloy

Laser controlled melting of pre-prepared Ti-6Al-4V surface is carried out. The alloy surface was initially coated with a carbon film using a water soluble phenolic resin and TiC powders are imbedded in the coating prior to the laser treatment process. The discussions on the findings are presented in line with the previous study [3].

Figures 6.11 show optical and SEM micrographs of the laser treated surface. The presence of the laser scanning trackers is evident from the SEM micrograph. The overlapping ratio of the irradiated spot is in the order of 60 %, which appears as

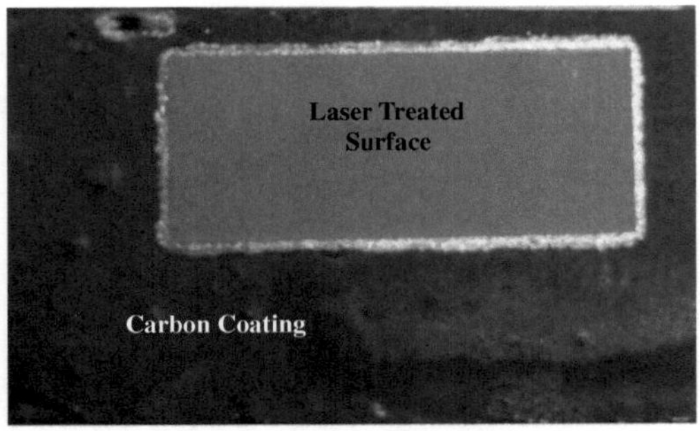

Laser Treated Workpiece (Optical Micrograph)

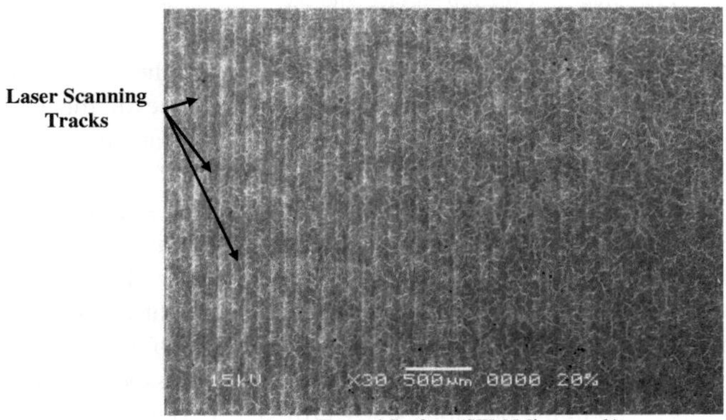

Laser Treated Surface (SEM Micrograph)

Fig. 6.11 Optical and SEM micrographs of the laser treated workpiece surface

continuous melting of the surface in the micrograph. In addition, no cracks and surface defects, such as voids and large cavities are observed. It should be noted that the large cavities are developed because of the evaporation of the surface during the excessive heating at high laser power intensity. The crack free surface reveals that thermal stress developed in the surface region does not form cracks in this region. In addition, the level of the micro-stress formed, due to the grain refinement at high cooling rates, is not significantly high to cause the crack initiation at the micro-level in the surface region. Moreover, during the multi-scanning of the surface by the laser beam acts as post-heat treatment process for initially scanned tracks while lowering the stress levels in the treated layer. The similar observation is reported in the previous study [19]. The close examination of the surface at high magnification, as observed from SEM micrograph, reveals that fine structures with nano-sized grains are formed in the surface vicinity. This occurs because of the followings: (1) formation of TiN and Ti(C, N) compounds causes volume shrinkage in the surface region resulting in dense structures [20] and (2) high cooling rates after the laser treatment process causes grain refinement in the surface region. It should be noted the formation of fine structures consist of nano-sized grains are measured in

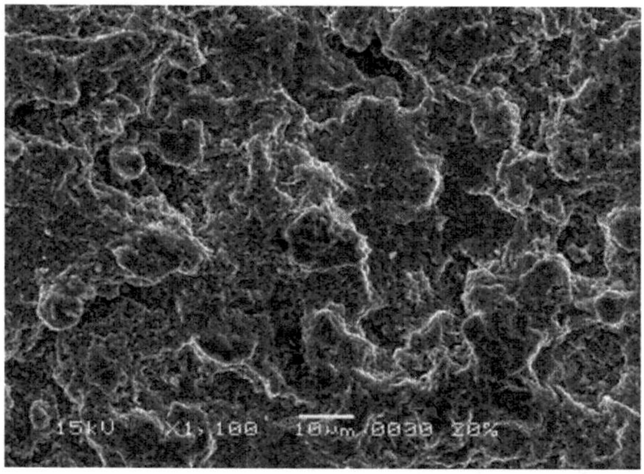

Fig. 6.12 SEM micrographs of the laser treated surface texture irregularities

the surface region of the laser nitrided titanium [21]. It should be noted that temperature at the surface attains considerably high values, since TiC appears to be almost dissolved totally in the surface region and TiC particles is not clearly apparent at the treated surface. The averaged roughness of the surface (Ra) is in the order of 4.5 μm, which is less than the TiC average particle size. This also indicates the melting and/or partial melting of TiC particles. In addition, the attainment of relatively high surface roughness value is associated with the high surface temperature during the treatment process. In this case, the melt flow under the high assisting gas pressure contributes to the enhancement of the surface roughness. The microscale surface texture irregularities due to the melt flow at high assisting gas pressure are also observed from SEM micrograph (Fig. 6.12).

Figure 6.13 shows SEM micrographs of the cross-sectional view of the laser treated surface. The fine structures are evident in the surface region of the laser treated layer. However, fine dendrites are formed in the vicinity of the surface indicating nitrogen-rich species in this region [22]. In addition, feathery-like structures are also observed in this region, which is also associated with the presence of nitrogen-rich nitride compounds. Moreover, the volume shrinkage due to the formation of dense structure in the surface region results in locally scattered few microsize voids, which is observed in SEM micrograph. However, partially melted TiC particles embedded in the surface region is also observed. This appears to be few in number and mainly located in the vicinity of the surface. Moreover, a dense layer is formed around the partially melted particle. This indicates the formation of nitride species in the molten part of the particle as well as its surrounding. Although the thermal expansion coefficient of nitride species and TiC are different, no micro-crack is observed around the partially melted particles. The elongated dendrites are observed at some distance from the treated surface towards the solid bulk. The presence of the elongated dendrites reveals the formation of titanium nitride compounds in this region. The possibility of forming nitrocarbide compounds in this region is less likely, since they are associated with small dendrites or fine structures. As the depth below the treated surface increases

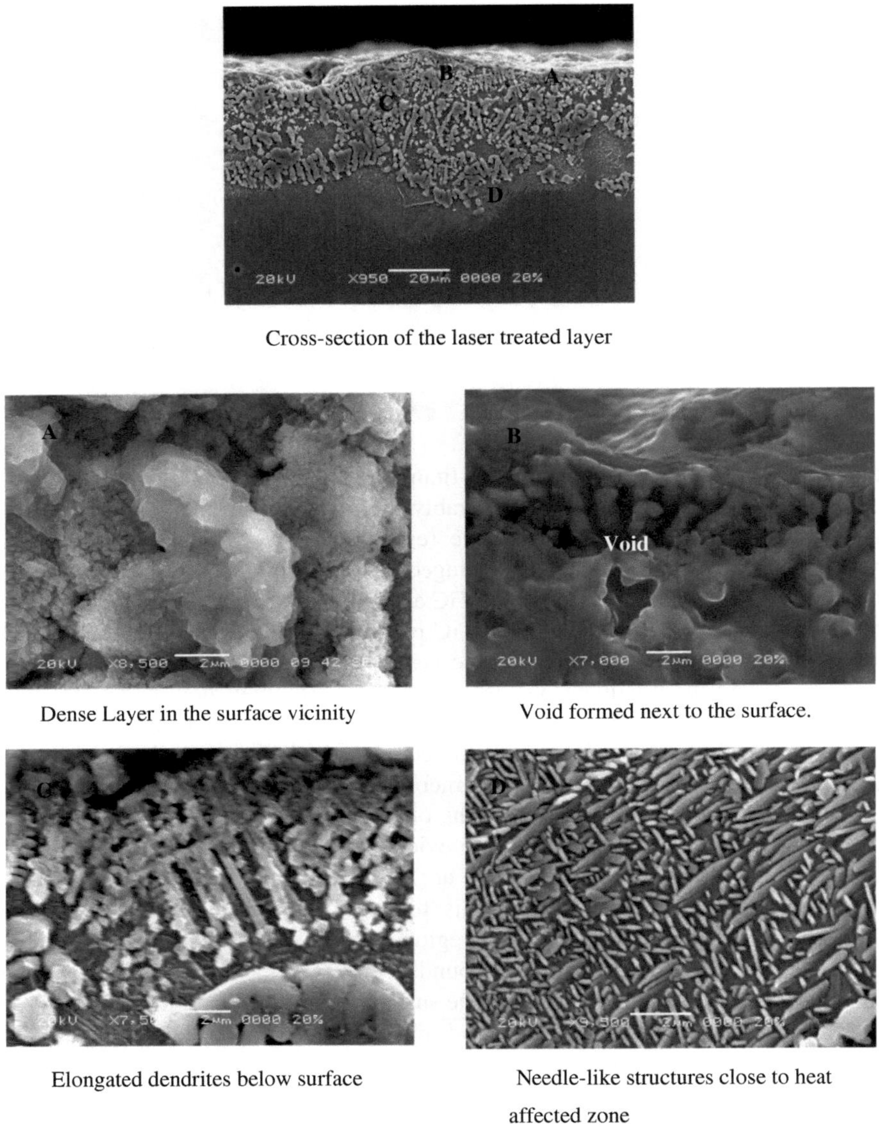

Fig. 6.13 SEM micrographs of laser treated layer cross-section. **a** Dense layer in the surface vicinity, **b** void formed next to the surface, **c** elongated dendrites below surface, **d** needle-like structures close to heat affected zone

further, the short needle-like structures are observed. This shows the presence of α'-Ti/TiN_x in this region [19]. However, as the distance increases further, small needle-like structures replaces with large needle-like structures, which is

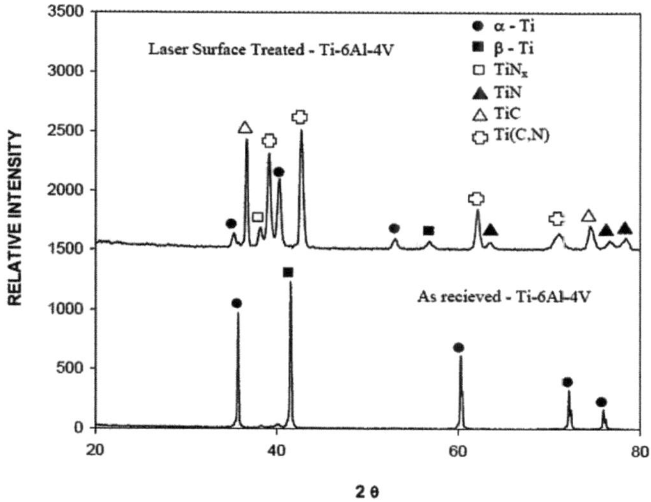

Fig. 6.14 XRD diffractogram for as received Ti-6Al-4V alloy and laser treated surface

associated with the heat affected zone. The size of heat affected zone is narrow because of the low thermal diffusivity of the alloy which suppresses the heat conduction in the heat affected zone.

Figure 6.14 shows XRD diffractograms of as received and laser treated surfaces. It is evident that TiC, TiN_x, Ti_2N, TiC_xN_{1-x} peaks is visible in the diffractogram. It should be noted that TiN, Ti(C,N), TiC are identified in accordance with ICDD 38-1420 card, ICDD 42-1488 card, and ICDD 32-1383 card. The presence of TiC peaks reveals that partially dissolved. TiC particles are present in the surface region. In addition, the formation of new TiC compounds at elevated temperature is also possible due to the presence of carbon film at the workpice surface, which is formed prior to the laser treatment process. The presence of TiN_x is associated with the fast solidification rate, which leads to a non-equilibrium state due to the time shortage for the interaction between Ti and N to form stoichiometric TiN in the surface region. Moreover, TiN_2 peak is associated with the formation of metastable TiN_x compound. The metastable phase is formed because of high cooling rates resulting in non-equilibrium process [19]. Due to the gradual heating resulted from the laser scanning of the neighboring tracks, the dissolution of metastable titanium nitride and the diffusion of nitrogen from the supersaturated titanium would give rise to formation of Ti_2N compound in the surface region. This takes place according to $a\mathrm{TiN_x} \rightarrow b\mathrm{Ti_2N} + c\mathrm{Ti}$. Moreover, the presence of non-uniform nitrogen concentration at the surface is evident from EDS results, which is given in Table 6.12. This is attributed to formation of different nitride compounds at the surface as well as non-equilibrium process taking place during non-uniform cooling of the surface.

Table 6.12 EDS spectrums (wt %) and their locations at the surface

	N	Al	Ti	V
Spectrum 1	5.61	1.94	89.21	3.24
Spectrum 2	4.03	1.84	91.60	2.53
Spectrum 3	9.39	3.71	84.76	2.15
Spectrum 4	9.89	2.97	84.59	2.56

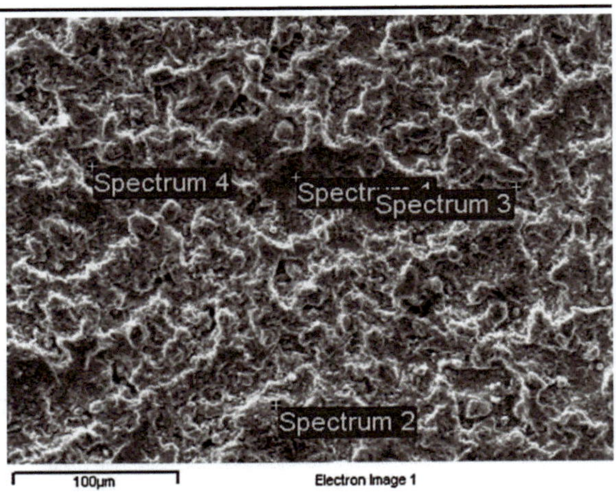

Table 6.13 Microhardness test results at the surface

	Microhardness (HV)
As received surface	305
Present study	1,650 ± 40
Laser nitrided [21, 23]	1,260 ± 30
Plasma nitrided [25]	1,050
Plasma nitrided [24]	925

Table 6.14 The residual stress predicted and measured at different depths inside the nitride layer

	Predictions (GPa)	XED technique (GPa)
Present study	−1.63	−1.55 ± 0.30
(Ti, Al) N coating [26]	–	−3.11
Laser Nitrided Ti–6Al–4V [25]	–	0.265

Table 6.13 gives the microhardness results obtained from the present study and the previous studies in relation to laser nitriding [23] and [24] as well as plasma nitriding of Ti-6Al-4V surfaces [25] and [26]. It should be noted that the microhardness across the treated layer could not be measured with accuracy, since the treated layer thickness extends 50 μm depth below the surface. The surface microhardness increases significantly as compared to the base material hardness. The increase in the hardness is associated with the formation of compact and fine structures in the surface region of the workpiece. In addition, the microhardness resulted from the carbonitrided surface is higher than those corresponding to the

laser nitrided surface [23] and [24], and the plasma nitrided surfaces [25] and [26]. Table 6.14 gives the residual stress predicted using the analytical formula developed previously [11] and from the XRD technique. In addition, the residual stresses determined from the XRD technique for laser assisted nitriding [27] and (Ti, Al) N hard coating [28] are also included for comparison. The residual stress predicted agrees with the measured data. However, the discrepancy between both results is associated with the assumption in the formulation, which considers the uniform structure at the surface. This may not be very true. It should be noted that the phase transformation during forming Ti_2N changes the stress states in the surface region [19]. Consequently gradual heating during the laser scanning of the neighboring tracks reduces the residual stress formed in the surface region.

References

1. Yilbas, B.S., Arif, A.F.M., Shuja, S.Z., Gondal, M.A., Shirokof, J.: Investigation into laser shock processing. ASM J. Mater. Eng. Perform. **13**, 47–54 (2004)
2. Yilbas, B.S., Arif, A.F.M., Karatas, C., Abdul Aleem, B.J., Tabet, N.: Laser gas-assisted nitriding of steel: residual stress. Ind Lubr Tribol **62**(4), 214–223 (2010)
3. Yilbas, B.S., Akhtar, S.S., Karatas, C.: Laser gas-assisted processing of carbon coated and TiC embedded Ti-6Al-4 V alloy surface. Appl. Surf. Sci. **257**(2), 531–537 (2010)
4. Yilbas, B.S., Sahin, A., Davies, R.: Laser heating mechanism including evaporation process initiating the laser drilling. Int. J. Mach. Tools Manuf. **35**(7), 1047–1062 (1995)
5. Yilbas, B.S., Sami, M.: Liquid ejection and possible nucleate boiling mechanisms in relation to laser drilling process. J. Phys. Part D: J. Appl. Phys. **30**, 1996–2005 (1997)
6. Peyre, P., Fabbro, R., Berthe, L., Dubouchet, C.: Laser shock processing of materials, physical processes involved and examples of applications. J. Laser Appl. **8**, 135–141 (1996)
7. Xiao, L.: Numerical computation of stress waves in solids. Akademie Verlag GmbH, Berlin (1996)
8. Smith, G.D.: Numerical solution of partial differential equations: finite difference methods, 3rd edn. Clarendon Press, Oxford (1985)
9. Gudonov, S.K.: A finite difference method for the numerical computation of discontinuous solutions of equations of fluid dynamics. Mat. Sb., pp. 271–306 (1959)
10. Ochoa, E.A., Figueroa, C.A., Alvarez, F.: Nitriding of AISI 4140 steel by a low energy broad ion source. J. Vac. Sci, Technol. A **24**, 2113–2116 (2006)
11. Stokes, J., Looney, L.: Residual stress in HVOF thermally sprayed thick deposit. Surf. Coat. Technol. **177–178**, 18–23 (2004)
12. Oliver, W.C., Pharr, G.M.: Improved technique for determining hardness and elastic modulus using load and displacement sensing indentation experiment. J. Mater. Res. **7**, 1564–1580 (1992)
13. Niihara, K., Morena, R., Hasselman, D.P.H.: Indentation fracture toughness of brittle materials for palmqvist crack. Fract. Mech. Ceram. **5**, 97–105 (1983)
14. Li, H., Khor, K.A., Cheang, P.: Young's modulus and fracture toughness determination of high velocity oxy-fuel-sprayed bioceramic coatings. Surf. Coat. Technol. **155**, 21–32 (2002)
15. Mao, K., Sun, Y., Bell, T.: An initial approach to the link of multi-layer coatings contact stresses and the surface engineered gears. Surf. Coat. Technol. **201**, 5796–5803 (2007)
16. Chu, J.P., Rigsbee, J.M., Banas, G., Elsayed-Ali, H.E.: Laser-shock processing effects on surface microstructure and mechanical properties of low carbon steel. Mater. Sci. Eng. **260A**, 260–268 (1999)

17. Watkins, T.R., England, R.D., Klepser, C., Jayaraman, N.: Measurement and analysis of residual stresses in-phase iron nitride layers as a function of depth, International Center for Diffraction Data 2000. Adv. X-Ray Anal. **43**, 31–38 (2000)
18. Pellizzari, M., Molinari, A., Straffelini, G.: Thermal fatigue resistance of gas and plasma nitrided 41CrAlMo7 steel. Mater. Sci. Eng., A **352**, 186–194 (2003)
19. Xin, H., Hut, C., Baker, T.N.: Microstructural assessment of laser nitrided Ti-6Al-4V alloy. J. Mater. Sci. **35**, 3373–3382 (2000)
20. Lackner, J.M., Waldhauser, W., Ebner, R., Bakker, R.J., Schoberl, T., Major, B.: Room temperature pulsed laser deposited (Ti, Al)CxN1-x coatings—chemical, structural, mechanical and tribological properties. Thin Solid Films **468**, 125–133 (2004)
21. Hoche, D., Schikora, H., Zutz, H., Queitsch, R., Emmel, A., Schaaf, P.: Microstructure of TiN coatings synthesized by direct pulsed Nd:YAG laser nitriding of titanium: development of grain size, microstrain, and grain orientation. Appl. Phys. **A 91**, 305–314 (2008)
22. Razavi, R.S., Salehi, M., Monirvaghefi, M., Mozafarinia, R.: Effect of laser gas nitriding on the microstructure and corrosion properties of Ti-6Al-4V alloy. ISIJ Int. **47**, 709–714 (2007)
23. Biswas, A., Li, L., Chatterjee, U.K., Manna, I., Dutta Majumdar, J.: Diode laser assisted surface nitriding of Ti-6Al-4V: properties of the nitrided surface. Metall. Mater. Trans. A **40A**, 3031–3037 (2009)
24. Yilbas, B.S., Karatas, C., Uslan, I., Keles, O., Usta, I.Y., Ahsan, M.: CO_2 laser gas assisted nitriding of Ti-6Al-4V alloy. Appl. Surf. Sci. **252**, 8557–8564 (2006)
25. Yilbas, B.S., Nickel, J., Coban, A., Sami, M., Shuja, S.Z., Aleem, A.: Laser melting of plasma nitrided Ti-6Al-4V alloy. Wear **212**, 140–149 (1997)
26. Fouquet, V., Pichon, L., Drouet, M., Straboni, A.: Plasma assisted nitridation of Ti-6Al-4V. Appl. Surf. Sci. **221**, 248–258 (2004)
27. Yilbas, B.S., Arif, A.F.M., Karatas, C.: Laser gas-assisted nitriding of Ti-6Al-4V alloy and residual stress analysis. Surf. Eng. **25**, 228–234 (2009)
28. Gobel, T., Menzel, S., Hecker, M., Bruckner, W., Wetzig, K., Genzel, C.: Stress measurements in thermal loaded (Ti, Al)N hard coatings. Surf. Coat. Technol. **142–144**, 861–867 (2001)

Chapter 7
Concluding Remarks

Absract Laser surface treatment is one of the effective methods in surface engineering because of the many advantages over the conventional techniques. However, the presence of high thermally induced stresses, due to high temperature gradients, local non-uniformity in microstructures, and defects in morphology, due to uncontrolled melting or/and evaporation, limit the practical application of lasers in surface engineering. In this chapter the findings related to all chapters are concluded in details according to appropriate sub-headings.

Laser surface treatment is one of the effective methods in surface engineering because of the many advantages over the conventional techniques. However, the presence of high thermally induced stresses, due to high temperature gradients, local non-uniformity in microstructures, and defects in morphology, due to uncontrolled melting or/and evaporation, limit the practical application of lasers in surface engineering. However, the proper selection of the operating parameters of lasers and prior preparation of the surfaces can minimize the limitations in the practical applications. In this book, analytical and numerical treatments of the laser surface processing are presented to provide physical insight for improved processing. In addition, some practical applications of laser surface treatment are included for the practical applications. Therefore, the conclusions derived from the present study are presented briefly under the appropriate sub-headings.

7.1 Analytical Treatment for Melting and Evaporation Processes

Although some useful assumptions are made for the analytical treatment of the surface heating problem, the closed form solutions provide functional relations between the variables and the parameters. The assumption of one-dimensional semi-infinite body can be justified by knowing that the axial heat conduction takes over the radial conduction due to the unidirectional absorption of the incident

B. S. Yilbas and S. Z. Shuja, *Laser Surface Processing and Model Studies*, Materials Forming, Machining and Tribology, DOI: 10.1007/978-3-642-36629-1_7,

beam (Beer Lambert's law) and considerably smaller absorption depth of the irradiated material as compared to its thickness.

In general, the pulse parameters and Biot number effect the surface temperature profiles when the repetitive pulse heating is introduced for the conduction limited heating case. In this case, increasing Biot number lowers the peak surface temperature for large values of Biot numbers. In addition, the convective cooling of the surface does not lower the peak surface temperature only, but influences the temporal behavior of the surface temperature. Energy balance attains among the internal energy gain, convection and conduction losses at some depth below the surface. The location at which the energy balance occurs changes with the pulse parameters. As the heating period increases further, the convection and conduction losses become important due to (1) the attainment of high surface temperature enhances the convective cooling of the surface, and (2) high temperature gradient accelerates the diffusional heat loss from the surface vicinity to the substrate bulk. In the case of non-conduction limited heating situations, the recession velocity obtained from constant temperature evaporation at the surface results in low magnitude of recession velocities at low power intensities, particularly in the early period of evaporation; however, it predicts substantially higher recession velocities for high laser power intensities. The equilibrium based recession velocity (steady heating situation) gives rise to high recession velocity at low laser power intensities, particularly for melting applications. In the case of surface evaporation, this suppresses surface temperature rise after initiation of the evaporation process. Therefore, surface temperature drops well below the evaporation temperature of the substrate material. Although evaporation initiation cools down the substrate surface, but temperature at the surface remains at around the evaporation temperature of the substrate material. Consequently, recession velocity determined from steady heating situation wrongly predicts surface temperature rise immediately after the evaporation process. Therefore, constant temperature evaporation at the surface estimates recession velocity correctly in the early evaporation period and recession velocity based on steady evaporation occurs as the evaporation period progresses further. In addition, there exists an evaporation period when recession velocities obtained from the constant temperature evaporation at the surface and steady heating situation become almost identical.

7.2 Numerical Treatment for Melting and Evaporation Processes

Numerical approach for the solution of the laser induced melting and evaporation is fruitful, since it provides almost the complete picture in terms of the physical process. However, the limitations on the type of grid used and the size of the solution domain incorporated in the analysis are the constraints one has to face in the numerical solutions. This is particularly true for the solution of the transient multi-physics problems such as laser induced laser melting and evaporation at the surface.

Temperature increase is sharper along the x-axis than that corresponding to the r-axis. This results in the high temperature gradient along the depth in the surface region. Since the temperature gradient is high and the high temperature region does not extend considerably inside substrate material, the attainment of steady temperature across the phase change is not observed from the figures. The phase change results in local peak in the temperature gradient because of the low temperature gradient and the extension of high temperature region along the r-axis (in the radial direction). The extension of the high temperature region and the melt pool size along the r-axis is associated with the laser power intensity distribution at the surface, which is Gaussian. The circulation cells are formed in the melt pool due to the Marangoni effect. However, the location of the center of the circulation cell moves towards the free surface as well as the edge of the pool. This is because of the thick shear layer developed above the mushy zone and the presence of high surface tension towards the melt pool circumference. The slip boundary between the mushy zone and melt layer interface is responsible for the formation of the thick shear layer above the mushy zone, provided that the shear resistance is low due to attainment of the low rate of fluid strain in this region. The maximum velocity gradient increases slowly during the initial formation of the melt pool. Once the circulation initiates in the melt pool, the maximum velocity magnitude increases with processing time at which the melt depth enhances. The predictions of the melt pool size agree with its counterpart obtained from the predictions.

In the case pulse repetition, as the pulse repetition progresses, temperature rise becomes gradual; in which case conduction, convection and radiation losses from the irradiated region suppress temperature rise. The size of melting and mushy zones changes at the pulse beginning and ending, which is more pronounced for the pulse repetitions onset of the melting. This replaces with a small change in mushy zone size for the late pulse repetitions. The influence of duty cycle on the melting process is significant; in which case, increasing pulse duty cycle results in deep and wide melting zones. In addition, temperature at the surface increases substantially. The size of mushy zone in the radial direction becomes larger for low duty cycles than high duty cycles. Temperature distribution in the radial direction does not follow the laser pulse intensity distribution in the same direction. This is because of the latent heat of melting, which modifies temperature distribution in this direction once the melting temperature is reached.

The laser intensity parameter (β) has significant effect on temperature distribution. In this case, the peak temperature reduces with increasing intensity parameter in such a way that the melting replaces with the solid phase heating as the intensity parameter increases beyond $\beta = 0.6$. The melting of the surface initiates earlier for $\beta = 0$ as compared to its counterparts corresponding to $\beta > 0$. As the intensity parameter increases, temperature difference between the heating and cooling periods of each pulse reduces. In addition, some small variation is observed around the melting temperature for increased intensity parameter. This is attributed to the melting in the heating period and solidification of the melt pool in the cooling period of the laser heating pulse. The counter rotating circulation cells are formed in the melt pool for intensity parameter other than zero, which is attributed to the melt size, since it becomes shallow for $0.4 \leq \beta \leq 0.6$.

The super heating of liquid phase occurs in the irradiated region and temperature decay is sharp in the region next to the liquid phase boundary. However, as the distance increases from the liquid boundary, temperature decay becomes gradual. The sharp decay in temperature results in high rate of heat flux from high temperature region to low temperature region. This, in turn, suppresses the enlargement of melt zone while increasing the size of heat affected zone. This situation is observed at low workpiece speed; in which case temperature attains the maximum at the irradiated zone as compared to its counterparts corresponding to the high scanning speeds. The depth of the melt zone extends almost 90 μm below the surface for workpiece speed of 0.05 m/s. However, increasing workpiece speed lowers the depth of the melt zone.

In the case of melting of two-materials, temperature rise is fast at the alumina-steel interface and in the vicinity of the steel surface during the heating cycle. Steel undergoes the phase change after the second pulse while phase change in alumina occurs at a later heating period. As the pulses progresses, temperature rise and decay in the heating and cooling cycles of the pulses become almost the same; in which case, alumina coating and steel surface undergo a phase change during the heating cycle. In the cooling cycle, the solidification occurs in the alumina coating while liquid phase remains in the surface of steel. The solid phase heating is associated with high liquidus and solidus temperatures of alumina. The formation of mushy zone at the liquid–solid interface is not notable in temperature curves, which is due to the small mushy zone size. The melt pool size in steel is higher than that corresponding to alumina. A circulation cell is formed in alumina due to the Marangoni effect. However, the flow ceases in the melt pool of steel, since steel does not have a free surface and the density variation is small in steel melt pool.

In the case of coating located at the top surface of the workpiece, temperatures corresponding to boron and alumina are higher than that of tungsten carbide coating. This is associated with high thermal diffusivity and high absorption coefficient of tungsten carbide. Surface temperature increase is significant during the initial period of heating cycle of the laser pulse while in the cooling cycle temperature decay is sharp due to the cease of laser power. The influence of laser intensity distribution on temperature profiles is considerable, since radial distribution of surface temperature follows almost the laser intensity distribution at the workpiece surface. Boron and alumina coatings melt completely during the heating cycle of 10th pulse for laser intensity parameter $\beta = 0$. In addition, steel beneath the coating undergoes melting at the interface region. Although coating remains in the solid phase for tungsten carbide, underlying steel melts while forming a melt pool at the interface region. The size of the melt pool is influenced by the laser intensity parameter; in which case, the orientation of the melt pool changes from the irradiated spot center for the laser pulse parameter $\beta = 1$. The depth of the melt pool becomes shallow influencing the Marangoni flow in the melt pool. In general, a circulation cell is developed and the velocity magnitude in the melt pool attains higher values for the laser intensity parameter $\beta = 0$ than that corresponding to $\beta = 1$.

In the case of surface evaporation, cavity extension in the axial direction is much higher than that corresponding to the radial direction due to laser power intensity distribution at the surface, which is Gaussian. The size of mushy zone generated across the liquid–vapor interface is smaller than that corresponding to vapor–liquid interface. This is because of the latent heat of evaporation, which is higher than the latent heat of melting. Velocity magnitude in the cavity attains high values in the early heating period due to rapid evaporation of the surface. Moreover, velocity magnitude attains high values in the region of the symmetry axis. Due to the momentum change between the vapor front and its ambient, which is water, a low velocity and high pressure region is generated along the symmetry axis in the vapor jet front. Continuous vapor generation, because of evaporation of surface, enables to generate a high velocity region between the vapor front and cavity wall surface. This, in turn, results in complex flow structure in the cavity. This situation alters with time and secondary pressure peak is generated at the cavity exit as the heating progresses. Recoil pressure in the order of 10^{10} Pa is generated in the cavity due to high evaporating front velocity and relatively low recession velocity of the cavity surface. Moreover, recoil pressure remains high in the cavity with progressing time.

7.3 Practical Applications

The practical applications of laser surface treatment spans in wide areas of surface engineering. It includes, shock hardening, cladding, annealing, coating, particle injection, etc. The conclusions due to laser shock hardening, gas assisted nitriding, and control melting of prepared surfaces are presented herein.

In the case of laser shock hardening, microhardness as high as 1.8 times of the base hardness of the substrate material is achieved at the ablated surface. The hardening of the workpiece in the plastic region is because of the dislocation generated due to the propagating of the high amplitude plastic wave. SEM micrographs prevail that evaporation is the sole mechanism governing the mass removal from the surface during the ablation process. In this case, recoil pressure in the order of 0.8 GPa is generated, which in turn initiates the generation and propagation of the plastic waves into the substrate material. Once the recoil pressure diminishes due to ending of the evaporation process, unloading wave with higher speed than the plastic wave initiates and propagates into the workpiece. The time when these waves interact, the plastic wave diminishes. In the present case, the time at which two waves interact is in the order of 1.2×10^{-7} s and the depth where they interact is in the order of 400 μm.

In case of laser gas assisted nitriding of steel, the nitride compounds formed within the surface region of the substrate material is $\varepsilon - Fe_{2-3}N$, while as the depth below the surface increases domination of Fe_4N compound is observed. This is because of the nitrogen diffusion rate during the laser processing; in which case, nitrogen-rich compound are formed in the surface region. SEM micrograph reveals that nitrided region extended almost more than 60 μm below the surface and the nitrided layer is free from the microcracks and the voids. In the case of numerical study, temperature decays sharply immediately after the cooling cycle starts. This, in turn, results in high temperature gradients and high stress fields in the laser irradiated region. Moreover, towards the end of the cooling cycle (when temperature reduces to almost room temperature) the stress field becomes the residual stress in the substrate material. The residual stress predicted attains values within 230 MPa, which remains almost uniform in the nitrided layer, except in the surface region. The residual stress predicted is tensile, which is similar to that reported in the literature. Moreover, the residual stress reduces slightly due to the low temperature gradient developed in this region and the unconstrained expansion of the free surface. When comparing the residual stress predicted with the measurement results as obtained from the XRD technique as well as the indentation tests, all the results are in reasonably good agreement. The small discrepancies between the experimental data and predictions are attributed to the assumptions made in the model study and the measurement errors.

In the case of laser treatment of pre-prepared surfaces, the laser treated surface is free from surface irregularities such as cavities and cracks. Due to the high overlapping ratio of the irradiated spot at the surface, the laser scanning tracks appear as continuous melting sites. The melt flow, because of high temperature heating at the surface, and the partial dissolution of TiC particles increase the average surface roughness of the treated layer. The carbon coating and TiC embedded in the coating facilitate the formation of TiC_xN_{1-x}, which contributed to surface hardness and the formation of the compact and dense layer in the surface region of the treated layer. This results in a few scattered microsized voids in the surface vicinity of the treated layer. The small dendrites and feathery-like structures are formed in the surface region, which reveal the presence of nitride species in this region. As the depth increases below the surface needle-like structures are formed. The size of the heat affected zone is narrow because of the low thermal diffusivity of the titanium alloy. TiC_xN_{1-x}, TiN, TiN_x, TiC, and Ti_2N compounds are formed in the surface region, which contribute to the enhancement of the surface hardness of the treated layer. The microhardness increases significantly after the laser treatment process; however, non-uniform formation of the nitride reach compounds in the surfaces vicinity alters the microhardness at the treated surface.

Index

A
Absorption, 1–3, 5–7, 13, 15, 16, 18, 29, 30, 35, 39, 41, 55, 61, 70–73, 92, 94, 102, 106, 113, 119, 127, 139, 140, 142
Assisting gas, 5, 7, 11, 15, 16, 20, 24, 32–34, 41, 121, 123, 133

B
Beer–Lambert's law, 29, 39
Biot number, 18–20, 140
Boiling temperature, 81, 93, 94, 99, 116
Boussinesq approximation, 37
Buoyancy, 37, 38, 44

C
Carbide particles, 69, 70, 72, 75, 76, 133, 142
Carbon film, 123, 131, 135
Carman–Konscy equation, 36
Cavity, 3, 47, 91, 92, 94, 95, 97–101, 104–110, 131, 143
Coating, 59, 60, 62, 63, 65, 66, 68–79, 112, 121, 123, 142, 144
Concentration, 120, 121, 127–130, 135
Conduction loss, 46, 102, 140
Convection current, 34, 49
Convergence criterion, 38, 40, 43, 62
Cooling periods, 63, 64, 77, 141

D
Darcy law, 36
Dawson's integral, 11, 102
Diffusion coefficient, 96, 120, 129, 130
Dislocation, 112, 118, 127, 143
Displacement, 115
Duty cycle, 13, 24–27, 63, 70, 73, 141

E
Elastic–plastic material, 115
Elastic waves, 112, 125
Energy balance, 18, 27, 30, 32, 43, 93, 113, 140
Enthalpy equation, 34, 37, 40
Enthalpy porosity technique, 36, 42
Evaporation, 29, 33, 34, 43, 47, 75, 81, 82, 91–94, 101–103, 105, 106, 109, 112–114, 125, 131, 132, 139, 140
Exponential pulse, 6, 7, 16

F
Flow field, 34, 37, 41, 44, 63, 68, 69, 75, 103
Fluid strain, 47, 68, 141
Fracture toughness, 121, 122, 128
Fresnel law, 3
Full pulse, 6, 7, 11, 16, 19, 20

G
Gas assisted, 32, 111, 112, 118, 119, 123, 127, 130, 143
Gaussian intensity, 13, 50, 51, 64, 75, 92
Grid independence, 15, 38, 40, 42, 43, 62, 100, 101

B. S. Yilbas and S. Z. Shuja, *Laser Surface Processing and Model Studies*, Materials Forming, Machining and Tribology, DOI: 10.1007/978-3-642-36629-1,

H

Heat transfer coefficient, 15, 17, 20, 636, 38, 62, 119
Heating periods, 26, 27, 44, 66, 70–73, 75–78, 106, 109
Hooke's law, 115

I

Internal energy gain, 2, 17, 18, 20, 25, 27, 44–46, 50, 71, 72, 76, 102, 127, 140

L

Laser parameter, 63, 69
Latent heat, 31, 33, 36, 43, 44, 63, 66, 82, 94, 106, 112–114, 143
Latent heat of evaporation, 82, 94, 106, 112–114, 143
Latent heat of melting, 31, 36, 43, 44, 63, 66, 141, 143
Linearly elastic, 117
Liquid fraction, 36, 37, 40, 42, 62
Liquid layer velocity, 34
Liquid phase, 3, 30, 38, 49, 51, 53, 54, 57, 64–66, 91, 105, 142

M

Machining intensity, 3, 81
Marangoni flow, 3, 34, 41, 48, 49, 51, 59, 60, 68, 142
Mass fraction, 93, 98, 99, 101
Maximum melt depth, 41, 55, 66, 68, 77, 79, 141
Maxwell's law, 113
Melt depth, 41, 55, 66, 68, 77, 79
Melting, 2, 3, 29–34, 36, 41, 43, 44, 47, 53, 54, 57, 59, 60, 62–66, 69–73, 75–77, 79, 81, 82, 93, 94, 104, 112, 114, 118, 123, 131–133, 139–144
Melt layer thickness, 32
Melt pool, 2, 3, 34, 35, 38, 41, 44–52, 54, 56, 57, 59, 63, 68, 69, 75, 76, 79, 141–143
Microhardness, 118, 122, 124, 126, 136, 143, 144
Molten flow, 233
Momentum sink, 36, 37
Moving heat source, 31, 52
Mushy zone, 3, 34, 36, 37, 42, 44, 45, 47, 50, 51, 57, 62, 65–68, 71–73, 75, 79, 92, 94, 100, 106, 141–143

N

Natural convection, 38, 93, 119
Navier–Stokes equation, 37
Nitriding, 111, 112, 118, 120, 121, 123, 127, 128, 130, 131, 136, 137, 143
Normal stress, 115

P

Phase change, 1, 2, 29, 34, 36–39, 41, 42, 44, 53, 60, 62–65, 73, 75, 77, 79, 81, 82, 93, 99–101, 103–105, 141, 142
Plastic deformation, 112, 114, 125
Plastic wave, 112–115, 125, 143
Pulse fall time, 49, 61, 63, 70
Pulse heating, 2, 6, 11, 13, 16, 20, 52, 63, 140
Pulse length, 15, 40, 49, 61, 63, 70, 81, 104, 117, 125
Pulse parameter, 6, 11, 16, 17, 20, 39, 49, 52, 57, 63, 70, 75–79, 82, 101–103, 140, 142
Pulse profile, 11, 19, 39, 48
Pulse repetition, 63, 141
Pulse rise time, 49, 61, 63, 70

Q

Q-switched pulses, 112
Quality of liquid, 93

R

Rate of dissipation, 96
Recession velocity, 30, 32, 43, 82, 91, 98, 99, 106, 108, 113, 140, 143
Repetitive pulses, 6, 12, 13, 15, 20, 27, 39, 61
Residual stress, 118, 119, 121, 127, 128, 136, 137, 144
Resolidification, 48

S

Schmidt number, 96
Shear force, 33
Shock processing, 106, 111–113, 117, 124
Single pulse, 13
Solid phase, 3, 5, 33, 35, 36, 44, 53, 56, 63, 65–67, 71, 73, 79, 92, 105, 114, 141, 142
Source term, 13, 16, 19, 34, 37, 41, 92, 119
Step function, 12
Step input pulse, 39

Strain, 47, 68, 115, 141
Surface reflectivity, 30, 35, 41, 61, 68, 82, 119
Surface temperature, 10–13, 17, 18, 20–22, 24, 27, 31, 32, 43, 48–50, 69, 73, 74, 77, 78, 81, 90, 92, 101, 102, 133, 140, 142
Surface tension, 3, 38, 47, 59, 68, 141
Surface treatment, 1, 2, 30, 34, 41, 111, 112, 123, 124, 131, 139, 143

T
Temperature gradient, 25, 27, 30, 44–47, 49–51, 54, 56, 58, 70, 71, 73, 76, 102, 111, 128, 139–141, 144
Thermal barrier, 66
Thermal conductance, 34
Thermal expansion, 37, 44, 121, 133
Thermalization time, 1, 2, 5, 6, 30
Thermal stress, 111, 113, 132
Turbulence kinetic energy, 96, 99
Turbulence model, 94, 95, 103
Two-layer structure, 60, 62, 63
Two-pulses, 13, 53, 57, 59, 60, 61, 63

V
Vapor pressure, 114
Volume shrinkage, 48, 132, 133
Volumetric heat source, 2, 13, 29, 34, 39, 40, 41

W
Work-hardening, 115, 117

MIX
Papier aus verantwortungsvollen Quellen
Paper from responsible sources
FSC® C105338

If you have any concerns about our products,
you can contact us on
ProductSafety@springernature.com

In case Publisher is established outside the EU,
the EU authorized representative is:
Springer Nature Customer Service Center GmbH
Europaplatz 3, 69115 Heidelberg, Germany

Printed by Libri Plureos GmbH
in Hamburg, Germany